Modern Electronic and Electrical Drafting with Computers

Modern Electronic and Electrical Drafting with Computers

JAMES D. BETHUNE
Boston University, College of Engineering

PRENTICE HALL, Englewood Cliffs, NJ 07632

Library of Congress Cataloging-in-Publication Data

Bethune, James D.
    Modern electronic and electrical drafting with computers / James
D. Bethune.
      p. cm.
    Includes index.
    ISBN 0-13-593310-2
    1. Electronic drafting—Data processing. 2. Computer-aided
design. I. Title.
TK7866.B495 1990
621.381′022′1—dc20

89-33397
CIP

Editorial/production supervision and
  interior design: Ed Jones/Julie Boddorf
Cover design: Diane Saxe
Manufacturing buyer: Robert Anderson/Mike Woerner

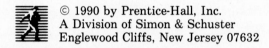 © 1990 by Prentice-Hall, Inc.
A Division of Simon & Schuster
Englewood Cliffs, New Jersey 07632

Printed in the United States of America
10 9 8 7 6 5 4 3 2 1

ISBN   0-13-593310-2

PRENTICE-HALL INTERNATIONAL (UK) LIMITED, *London*
PRENTICE-HALL OF AUSTRALIA PTY. LIMITED, *Sydney*
PRENTICE-HALL CANADA INC., *Toronto*
PRENTICE-HALL HISPANOAMERICANA, S.A., *Mexico*
PRENTICE-HALL OF INDIA PRIVATE LIMITED, *New Delhi*
PRENTICE-HALL OF JAPAN, INC., *Tokyo*
SIMON & SCHUSTER ASIA PTE., LTD., *Singapore*
EDITORA PRENTICE-HALL DO BRASIL, LTDA., *Rio de Janeiro*

To My Family

Contents

3

Two-Dimensional Constructions   29

4

Electronic Symbols   59

5

Schematic Diagrams   74

6

Printed Circuit Diagrams   95

7

Block and Logic Diagrams   133

8

Integrated Circuits   152

9

Orthographic Views and Chassis Drawings   172

10
Connection Diagrams   206

11
Residential Electrical Wiring   223

12

Industrial Wiring Diagrams   235

13

Graphs and Charts   256

Appendices 275

Index 286

Preface

This book explains how to prepare electronic and electrical drawings and schematics using CAD. It is a generic approach and does not rely on any single system. The book reviews some basic mathematical concepts, such as coordinate systems, that are needed to set up and prepare drawings using CAD. It explains two-dimensional constructions and shows how these techniques are applied to a variety of electronic and electrical drawings.

The book covers electronic symbols, schematic diagrams, and printed circuit development, including design layouts, film masters, soldering, and drill drawings. Integrated circuits are explained, and there are sample problems that show how to draw the various masks required for IC manufacturing.

Orthographic views are covered and applied to chassis drawings. There is a section on sheet metal bending and dimensioning of multiple-hole patterns.

There is a chapter on residential wiring, a chapter on connector diagrams, and a chapter on industrial wiring diagrams that includes one-line diagrams, ladder diagrams, and riser diagrams. The book concludes with a chapter on graphs and charts.

The book contains many sample problems that are explained in a step-by-step format accompanied by many illustrations. Each chapter contains exercise problems designed to apply the material presented in the chapter.

Several people deserve recognition for their help in the development of this book: Greg Burnell, my editor, for his constant encouragement and good marketing sense; Ed Jones and Julie Boddorf, the production editors, who put it all together; and Debbi Herman, Diane Cotter, and David George, who helped with everything from artwork to numbering pages and illustrations. Thanks to each of you.

James D. Bethune
Boston

1

Introduction

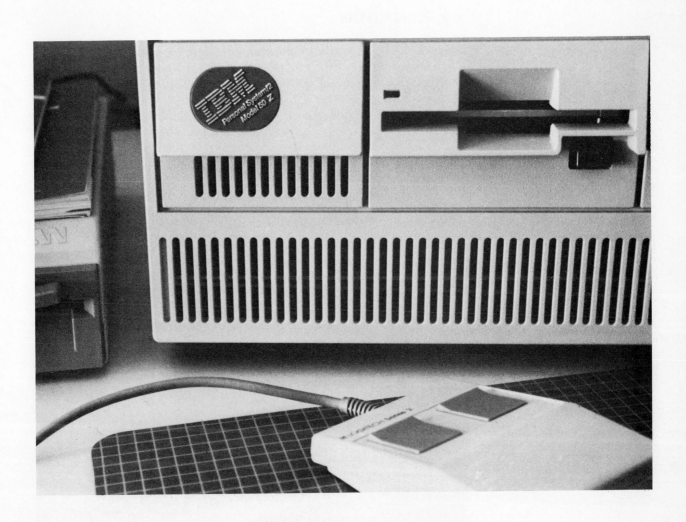

1-1 COMPUTER GRAPHIC SYSTEMS

Computer graphic systems are made from several components: a computer, a screen, a keyboard, and a device to input to the screen, such as a light pen, mouse, or thumbscrews. The size, capacity, and accuracy of these individual components determine the overall quality and capabilities of the system. Figure 1-1 shows a typical computer graphic system.

1-2 COMPUTERS

The principal component of any computer graphic system is a *computer*. The computer drives the entire system. Its memory capacity, speed of processing, and overall size will determine the capability and speed of the system.

Computers are, in general, rated by their *memory* capacity. Memory can either be *permanent*, called ROM (read-only memory), or *intermediate*, called RAM (random access memory). ROM is built into the computer and cannot be erased. A computer that has "built-in BASIC" has the fundamentals of the language BASIC stored in its ROM.

RAM memory is the memory used for most on-the-screen work. That is, as drawings are created, the computer is probably operating in its RAM mode. After a drawing is complete, it is transferred into a file and saved. It is important to mention that once the computer is turned

FIGURE 1-1 Computer graphics system. (Courtesy of Apple Computer Inc.)

off, that is, the power is off, the RAM will be wiped clean. Any work not filed correctly will disappear.

Computers derive their memory capacity from *chips*. Chips are pieces of silicon with integrated circuits comprised of thousands of small switches, transistors, resistors, and capacitors etched into their surface. A single chip about 1½ in. long and ½ in. wide will contain about 100,000 functions.

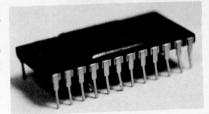

FIGURE 1-2 Computer chip.

Most computer companies do not manufacture their own chips; they purchase them from individual companies specializing in chip manufacture. The rivalry between chip manufacturers is intense. Each company continually tries to produce chips that are smaller, with larger capacity, that operate faster, and can be sold at a lower price. Figure 1-2 shows a computer chip.

Chips are combined into larger circuits called boards or *cards*. Figure 1-3 shows a board from a computer. One edge of a board is fitted with metal strips that slide into receptacles within the computer.

Many smaller computers can be upgraded — given greater memory or operating capabilities — by adding boards. For example, a "color board" will give a computer the ability to understand and process color commands.

Computer memory capacity is specified by the number of memory locations available. A 64K computer has 64,000 memory locations; a 256K has 256,000 locations. The term *Meg*, short for the prefix *mega*, meaning millions, is also used to identify memory capacity. A 40-Meg memory has 40,000,000 memory locations.

Graphics work in general requires large amounts of memory both for the software used to create the drawing packages and for the actual

FIGURE 1-3 Computer circuit board. (Courtesy of IGC Inc., 1568 Ocean Ave., Bohemia, NY 11716.)

FIGURE 1-4 Computer that stands on the floor near a workstation. (Courtesy of International Business Machines Corporation.)

drawings. Unfortunately, large memory capacity usually costs large amounts of money. However, as chips become more powerful, the ability of smaller systems to create more detailed and accurate graphics increases.

Systems with large memories require a space larger than is convenient for desktop computers. Figure 1-4 shows a computer stored in a small box located beside an operator's desk or workstation.

1-3 SCREENS

Computer *screens*, also called *visual display units* (VDUs), are rated by the number of *pixels* they contain. A pixel is like a small dot on the screen that is turned on or off to create the screen picture. The more pixels within a screen, the better the resolution (clarity) of the screen. A high-resolution screen might have 640 × 400 pixels, that is, a pattern consisting of 640 pixels from left to right of the screen and 400 pixels from top

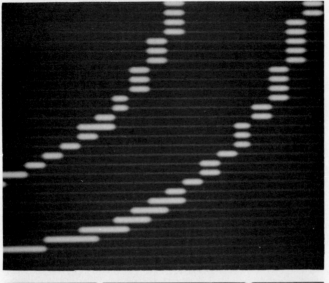

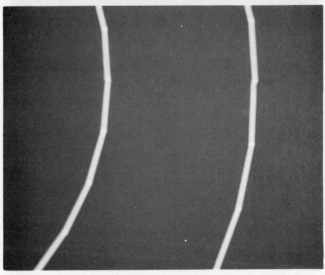

FIGURE 1-5 Comparison of high- and low-resolution screens.

to bottom. A low-resolution screen might have a pixel rating of 320 × 200. Figure 1-5 shows a comparison between the graphics produced by high-resolution and low-resolution screens.

1-4 INPUT DEVICES

In addition to keyboards, other devices are used to input information into the computer. The system shown in Figure 1-6 shows the use of a *light pen;* Figure 1-7, a *mouse* (in the operator's right hand), and Figure 1-8, *thumbscrews.* All these devices are used to control a screen cursor, which, in turn, is used to input information into the computer.

Another input device that is used in combination with a mouse or similar device is a *digitizing board.* Figure 1-9 shows a digitizing board. The digitizing board has small sensors embedded within its surface. A mouse is used to activate these sensors and to pass the location onto the computer. Shapes can be electronically traced and defined in numerical or digital information, which, in turn, can be processed by the computer.

FIGURE 1-6   Light pen.

FIGURE 1-7   Mouse. (Courtesy of International Business Machines Corporation.)

LIGHT PEN

MOUSE

THUMB SCREWS

FIGURE 1-8   Devices used to control the screen cursor.

FIGURE 1-9   Digitizing board.

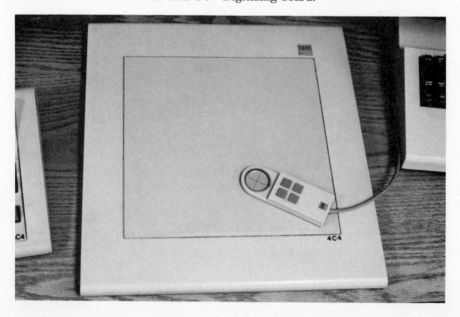

1-5 GLOSSARY OF COMPUTER GRAPHIC TERMS

Following is a glossary of terms commonly used when working with computer graphic systems.

Algorithm   Any special method used to solve a problem. The word usually refers to the mathematical equations used by a computer to perform specific functions.

Analog   A circuit that works on voltage values rather than numbers.

Batch Processing   Data processing that requires no interaction from the user.

Bit   The smallest piece of digital information.

Blanking   Erasing all information within a window.

Buffer   A location within the computer memory that serves as a temporary holding area until information is recalled.

Byte   Eight bits combine to form a byte.

Character   A number, letter, decimal point, punctuation mark, and so on. A normal line of written text has about 80 characters.

Chip   A piece of silicon that contains one or more integrated circuits.

Clipping   Erasure of all or part of an element.

Digital System   A system that handles information in discrete pieces such as numbers.

Element   A line or arc drawn on the screen.

EPROM   ROM that can be erased and/or programmed.

Error Message   Computer message to the user that appears on the screen and tells the user what error has been made.

Floppy Disk   A flexible disk that is used to store data. Information can be transferred from a floppy disk to the computer's RAM, or from the RAM to the floppy disk.

Gate   A simple on–off switch.

Hard Copy   The drawings or text printout produced by a computer system via a printer or plotter.

Hardware   The physical components that form a system: the computer, screen, printer, and so on.

Image Processing   The overall process of generating pictures on a computer.

Integrated Circuit   A series of very small circuits etched into a piece of silicon.

Interactive   A system that receives, processes, and returns data to a user. Systems that have good interactive capabilities are called "user friendly."

Local Area Network   A system that links together terminals or computers.

LSI (large-scale integration)   An LSI system contains about 1000 gates.

Memory   Computer circuits that retain information.

Microcomputer   Any small computer that contains only a few circuit boards. Some microcomputers contain only one chip.

Microprocessor   That part of the computer that processes digital information.

Mode   A point where two or more lines intersect.

Modem   A device that enables the computer to receive and transmit information over telephone lines.

Monochromatic Screen   A screen that displays only one color, usually white or green.

Number Crunching   A slang term that refers to a computer's ability to complete mathematical operations quickly.

On-Line   A computer or terminal that is in an operational mode.

Operator   A person using a computer.

Painting   Asking the computer to reproduce a known drawing on the screen.

Panning   Moving across a drawing that is larger than the screen.

Pixel   A single display element on a raster screen that can be varied in color and intensity.

Primitive   A basic command to the system. Usually, a single word that cannot be used for any other purpose. For example, if the word DRAW is a primitive within a given system, it cannot be used as a file name.

Program   A set of instructions used to command computer operations.

RAM (random access memory)   The temporary computer memory.

Raster Screen   A screen made with pixels that are activated by a horizontally scanning beam. A TV screen is a raster screen.

Resolution   A determination of the clarity and distinguishability of pictures on a screen.

ROM (read-only memory)   The permanent memory of a computer.

Software   The programs that control computer operations.

Time Sharing   A system whereby a large computer is linked to individual terminals that, in turn, share access to the main computer.

User   A person operating a computer.

VLSI (very large scale integration)   VLSI systems usually contain over 50,000 gates.

Window   A special rectangular area within a drawing. It is used to define a specific area of a drawing so that the area can be worked on.

Zooming   To isolate and enlarge a specific area of drawing.

2

Mathematical Concepts

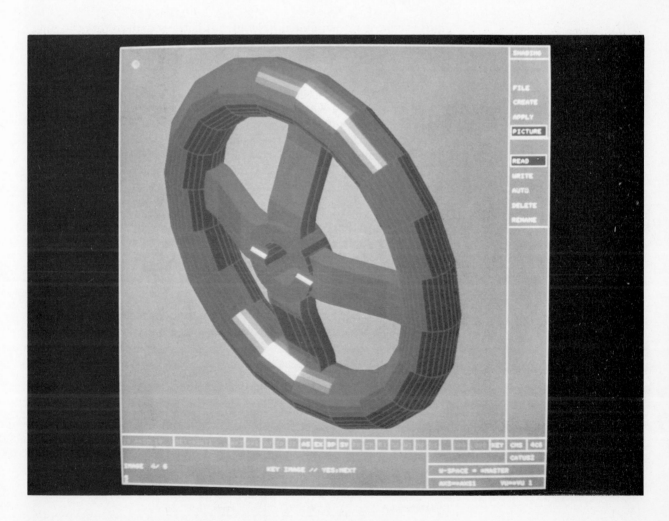

IBM 3270-PC/G Graphics Workstation. (Courtesy of International Business Machines Corporation.)

2-1 INTRODUCTION

In this chapter we introduce and review the mathematical concepts essential to the understanding and operation of computer graphic systems. English, metric, and screen units are covered together with six different types of coordinate systems: rectangular, rotated, polar, three-dimensional, and absolute versus relative.

2-2 UNITS

A *unit* is a basic measurement of length. It is the standard for other measurements. Technical drawing uses two basic units of measure, the millimeter as defined by the Système International d'Unités (SI units), and the inch as defined by U.S. Customary units.

The *metric system* is based on the meter. The meter is officially defined as 1,650,763.73 wavelengths of light as emitted from a transition state of a kryton-86 atom.

A meter is divided into larger and smaller units based on multiples of 10. This makes conversion between units easy. All that is required is a shift in the decimal point. The standard units of a meter are as follows.

$$
\begin{aligned}
0.001 \text{ meter} &= 1 \text{ millimeter} \\
0.01 \text{ meter} &= 1 \text{ centimeter} \\
0.1 \text{ meter} &= 1 \text{ decimeter} \\
10 \text{ meters} &= 1 \text{ decameter} \\
100 \text{ meters} &= 1 \text{ hectometer} \\
1000 \text{ meters} &= 1 \text{ kilometer}
\end{aligned}
$$

All measurements on technical drawings are made using millimeters. Figure 2-1 lists various subunits of a meter together with their conversion factors (multiples of a meter) and their abbreviations.

FIGURE 2-1 Multiples of a meter.

UNIT	Multiple of a meter	ABBREVIATION
meter	1	m
nanometer	10^{-9}	n n
micrometer	10^{-6}	μm
millimeter	10^{-3}	mm
centimeter	10^{-2}	cm
decimeter	10^{-1}	dm
decameter	10^{1}	dam
hectometer	10^{2}	hm
kilometer	10^{3}	Km
megameter	10^{6}	Mm

Technical drawings, done using English units, are based on an inch. An inch is defined as 0.0254 meter exactly. Small measurements can be specified using fractions of an inch (¼, ½, ¾, etc.) or decimal inches (0.25, 0.50, 0.75, etc.). Decimal inch-values are preferred.

Electronic drawings use unit values measured in ohms, farads, henrys, watts, volts, and amperes. Any unit value may be used with a prefix to change its value by an appropriate factor of 10. Figure 2-2 lists 14 prefixes that are used on electronic drawings. Also included is the symbol for each prefix and its equivalent multiple or submultiple.

Multipliers and submultipliers serve to shorten drawing notations and unit references. For example, one picofarad can be written 0.000 000 000 001 farad or 1×10^{-12} farad. Using the submultiplier for pico (P), one picofarad is written 1PF. One million tons can be written 1,000,000 tons or 1×10^{6} tons or using the multiplier *mega:* 1Mtons.

PREFIXES

Prefix	Symbol	Multiples and Submultiples
Tera	T	10^{12}
Giga	G	10^{9}
Mega	M	10^{6}
Kilo	K	10^{3}
Hecto	H	10^{2}
Deka	DA	10
Deci	D	10^{-1}
Centi	C	10^{-2}
Milli	m	10^{-3}
Micro	U	10^{-6}
Nano	N	10^{-9}
Pico	P	10^{-12}
Femto	F	10^{-15}
Atto	A	10^{-18}

FIGURE 2-2   Prefixes for multiples and submultiples.

Examples

2 Picofarads = 2PF = .000 000 000 002 F

4 Milliohms = 4mΩ = .004 Ω

6 Kilowatts = 6KW = 6000W

23.6×10^{4}	–	236,000
23.6×10^{3}	–	23,600
23.6×10^{2}	=	2,360
23.6×10^{1}	=	236
23.6	=	23.6
23.6×10^{1}	=	2.36
23.6×10^{2}	=	.236
23.6×10^{3}	=	.0236
23.6×10^{4}	=	.00236

COMPONENT	UNITS	SYMBOL
Resistor	Ohms	Ω
Capacitor	Farads	F
Inductor	Henrys	H
	Watts	W
	Volts	V
	Alternating Current	AC
	Direct Current	DC
	Amps	A

2-3 CONVERSIONS

Inches are converted to millimeters by multiplying the inch value by 25.4.

$$\text{inches} \times 25.4 = \text{millimeters}$$

For example,

$$3 \text{ in.} \times 25.4 = 76.2 \text{ mm}$$
$$0.50 \text{ in.} \times 25.4 = 12.7 \text{ mm}$$

Millimeters are converted to inches by dividing by 25.4.

$$\frac{\text{millimeters}}{25.4} = \text{inches}$$

For example,

$$\frac{15 \text{ mm}}{25.4} = 0.59 \text{ in.} \qquad \frac{31.5 \text{ mm}}{25.4} = 1.24 \text{ in.}$$

2-4 SCREEN UNITS

Computer graphic screens are divided into distances called *screen units*. The length of a screen unit is in turn derived from the screen's pixel size. (The word *pixel* derives from the term "picture element.") High-resolu-

FIGURE 2-3 A line, 50 pixels long, shown on a low-resolution screen.

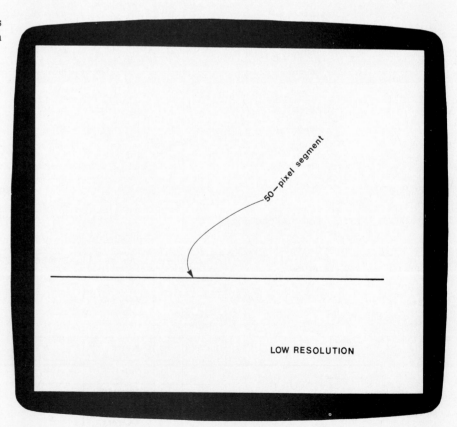

50–pixel segment

LOW RESOLUTION

tion screens can have 400 pixels vertically and 700 horizontally. Low-resolution screens can have as few as 40 pixels vertically and 60 horizontally.

Pixel size is not fixed, as with an inch or a millimeter, but varies according to the screen's capability. Figure 2-3 shows a line 50 pixels long drawn on a low-resolution screen. Figure 2-4 shows a line 50 pixels long drawn on a high-resolution screen. Note the difference in their lengths.

Pixel size will affect the clarity and accuracy of the picture presented on the screen. Figure 2-5 shows a circle drawn using a low-resolution screen, that is, one with large pixel units. Note the sharp edges. Figure 2-6 shows the same circle drawn on a high-resolution screen, that is, with very small pixel units. Note the smoothness of the curve.

Some computer systems have the ability to vary screen unit size by a viewing window command. This command changes the working size of the screen and allows the operator to work on small details or large overall drawings.

Some computer systems work exclusively in screen units and produce drawings and dimensions using screen units. Drawings done in screen units must be converted into either inches or millimeters if they are to be used as working drawings in manufacturing. An exception would be a system where a machine is linked directly to the computer graphic system, but even in this situation the screen units must be related to inches or millimeters for practical applications of the finished parts.

FIGURE 2-4   A line, 50 pixels long, shown on a high-resolution screen.

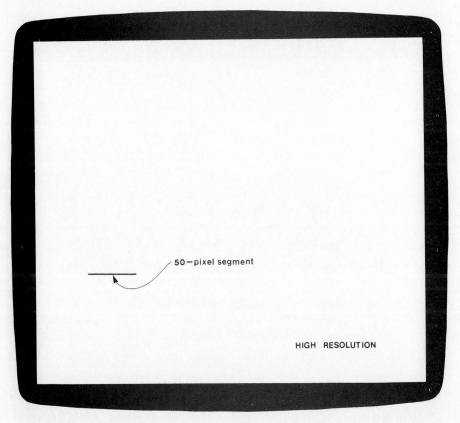

50—pixel segment

HIGH RESOLUTION

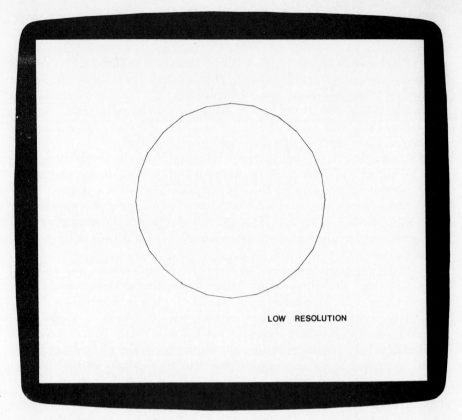

FIGURE 2-5 Circle drawn on
a low-resolution screen.

FIGURE 2-6 Circle drawn on
a high-resolution screen.

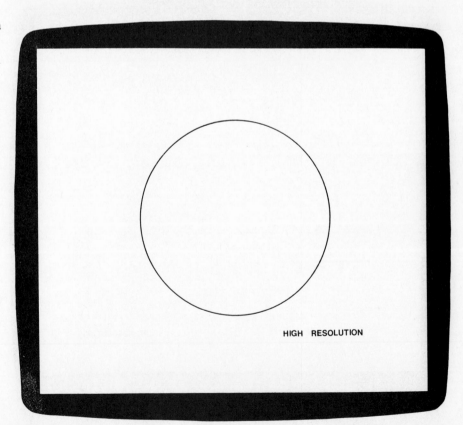

2-5 RECTANGULAR COORDINATE SYSTEM

A *rectangular coordinate system* is defined by two perpendicular lines, one horizontal and one vertical, that intersect at a point. The lines are called *axes*. The horizontal axis is the X *axis*, the vertical axis is the Y *axis*, and the point of intersection is the *origin* (Figure 2-7). A rectangular coordinate system uses both positive and negative X and Y values. X values located to the right of the origin are positive, X values located to the left of the origin are negative. Y values located above the origin are positive, Y values located below the origin are negative.

A rectangular coordinate system is divided into four *quadrants* (Figure 2-8). Each quadrant may stand alone as a separate coordinate system. For example, in Figure 2-9(a) the first quadrant is used to locate values with positive X and positive Y values. In Figure 2-9(b) the fourth quadrant is used to locate values with positive X and negative Y values. The second and third quadrants are for $(-X, +Y)$ and $(-X, -Y)$ values, respectively.

A location on a rectangular coordinate system is defined using an X and a Y value. The location is called a *point* and the X and Y values are the *coordinates* of the point. Coordinates are written in the form (X, Y). The first coordinate, the X value, is the horizontal distance from the origin. The second coordinate, the Y value, is the vertical distance from the origin. An X and a Y value are sufficient to locate any point in a rectangular coordinate system if the system is two-dimensional or planar.

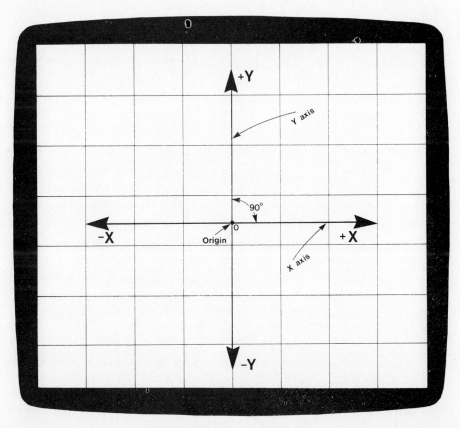

FIGURE 2-7 XY coordinate axis system.

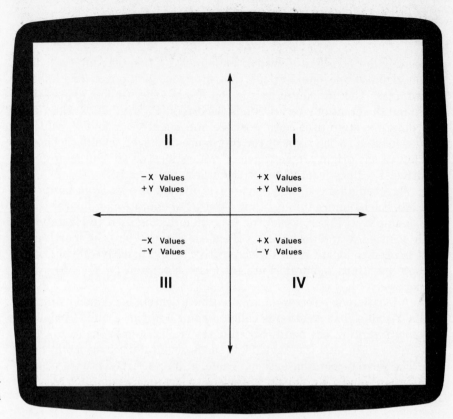

FIGURE 2-8 Coordinate values in the four quadrants of an *XY* coordinate system.

FIGURE 2-9 Each quadrant on an *XY* coordinate system may be used as a separate coordinate system.

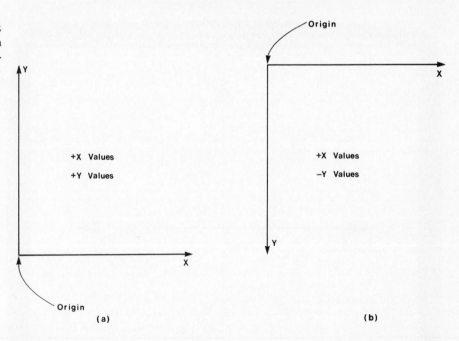

Figure 2-10 shows how to locate a point *A* whose coordinate values are (1, 4). In step 1, move horizontally 1 unit to the right of the origin. This defines the *X* value. In step 2, move vertically upward 4 units from the distance marked off in step 1. This defines the *Y* value. In step 3, mark the location of the point (1, 4). Define the point as *A*.

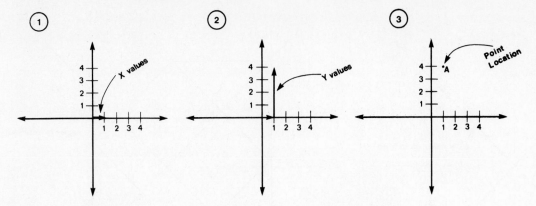

FIGURE 2-10   How to locate a point A whose coordinates are (1, 4).

Figure 2-11 shows how to find the coordinates of a point given its location relative to a known axis system. In step 1, count to the left from the origin along the horizontal axis to find the X coordinate value. (This is the negative X direction. The X value is -3.) In step 2, count upward from the origin along the vertical axis to find the Y coordinate value. The Y value is 2. It is acceptable to omit the positive sign when writing coordinates. A number with no sign stated is assumed positive.

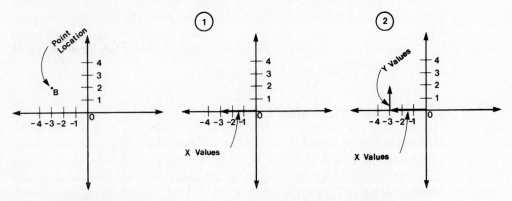

FIGURE 2-11   How to locate a point B whose coordinates are $(-3, 2)$.

2-6   ROTATED RECTANGULAR COORDINATE SYSTEM

A *rotated rectangular coordinate system* is a two-axis coordinate system turned through a specified angle. The 90° intersection of the two axes does not change. Figure 2-12 shows an axis system rotated 45°.

In computer-aided drafting, rotated coordinate systems are used for convenience. If the major portion of a drawing is constructed at an angle, it is easier to work with a rotated coordinate system than to reference values constantly back to a horizontal/vertical axis.

Figure 2-13 shows a figure constructed on a rotated axis system. The point coordinates relative to the rotated axis are $A(0,\ 2)$, $B(1,\ 1)$,

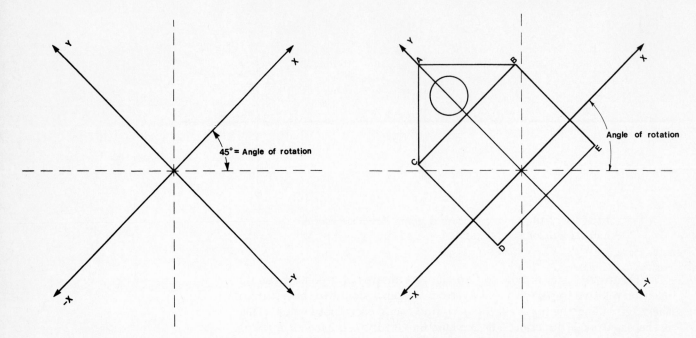

FIGURE 2-12 Rotated axis system.

FIGURE 2-13 Example of an object drawn on a rotated axis system.

$C(-1, 1)$, $D(-1, 1)$, and $E(1, -1)$. These values are measured directly from the rotated axis and need not be referenced to the horizontal/vertical axis.

Rotated rectangular coordinates are particularly useful when drawing auxiliary views.

2-7 POLAR COORDINATE SYSTEM

Polar coordinates locate a point using a radius and an angular value reference to an XY axis (Figure 2-14). Polar coordinates are most useful when working with circular or cylindrical shapes. They can also be used for dimensioning.

Polar coordinates are written in the form (r, ϕ), where r is the radius value and ϕ is the angular value. A polar coordinate of (12, 45) would

FIGURE 2-14 Polar coordinate system.

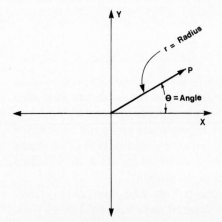

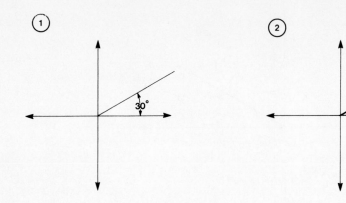

FIGURE 2-15   How to locate a point C whose polar coordinates are (5, 30).

mean a radius equal to 12 linear units and an angle of 45°. The linear units may be inches, millimeters, or screen units. The angular values are always measured in degrees.

Figure 2-15 shows how to locate a point C given the polar coordinates (5, 30), where 5 is defined in screen units. In step 1, a line 30° to the horizontal is drawn. In step 2, 5 screen units are measured along the 30° line defining the location of point C.

Negative polar coordinate values are interpreted differently by different computer graphic systems. Some systems permit only positive inputs; that is, they will not understand and process negative values. If point C of Figure 2-15 were to be located 30° below the horizontal, the polar coordinates using a positive-value-only system would have to be (5, 330). The counterclockwise direction is normally considered the positive angular direction; clockwise direction is negative.

Negative radius values are located in some systems by reflecting a negative value 180° from a positive value. For example, a polar coordinate of (−5, 30) would result in the reflected location shown in Figure 2-16, 180° from the positive 5 location. The actual degree value for the reflected point would be 210°.

Figure 2-17 shows another example of a point located using negative polar coordinate values. The value (5, 330) could be equivalent to the value (−5, 150) if the computer system reflects negative values.

FIGURE 2-16 Some computer graphic systems reflect negative linear values of polar coordinate inputs.

FIGURE 2-17 Two different coordinate values which locate a point in the same place.

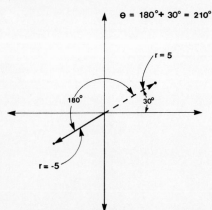

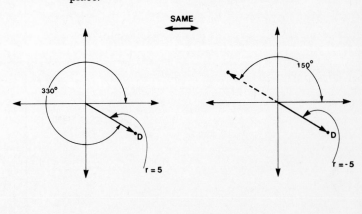

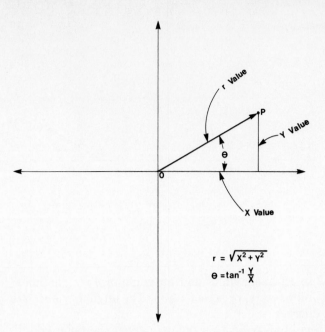

FIGURE 2-18 Formulas for converting between linear and polar coordinates.

$$r = \sqrt{X^2 + Y^2}$$

$$\theta = \tan^{-1} \frac{Y}{X}$$

Polar coordinate values can be changed to linear XY values using the formulas shown in Figure 2-18. If a computer system does not accept polar coordinates, it is sometimes possible to use these equations, stored in subroutines, to work with polar coordinate values by using their rectangular equivalents.

2-8 THREE-DIMENSIONAL COORDINATE SYSTEM

A three-dimensional rectangular coordinate system is defined by three axis lines: a horizontal (X), a vertical (Y), and a third line perpendicular to the plane formed by the horizontal and vertical lines (Z). All three axes lines intersect at 90° to each other. The point of intersection is called the origin (Figure 2-19). X values located to the right of the origin are posi-

FIGURE 2-19 Three-dimensional XYZ coordinate system.

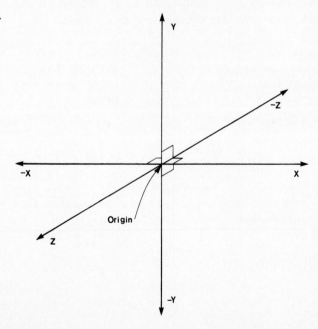

tive, X values located to the left of the origin are negative. Y values located above the origin are positive, Y values located below the origin are negative. Z values located in a direction out of the page are positive, Z values located in a direction into the page are negative.

Any point on a three-dimensional system is defined using three coordinate values. The three values are written in the form (X, Y, Z). The first coordinate is the horizontal distance from the origin, and the third is the distance "out of the page" from the origin.

Figure 2-20 shows how to locate a point, Q, whose coordinates are $(1, 2, 3)$. In step 1, move horizontally 1 unit to the right of the origin. This defines the X value. In step 2, move vertically upward 2 units from the distance marked off in step 1. This defines the Y axis. In step 3, move out of the page (parallel to the Z axis) 3 units. This defines the Z axis. Label the point $(1, 2, 3)$ by the letter Q.

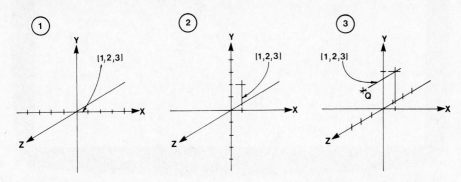

FIGURE 2-20   How to locate a point on a three-dimensional XYZ axis system whose coordinates are (1, 2, 3).

Figure 2-21 shows how to find the coordinates of a point given its location relative to the given axis. In step 1, count horizontally 4 units to the left from the origin to find the X coordinate. This is the negative X direction, so the X value equals -4. In step 2, count vertically upward 1 unit from the origin to find the Y coordinate. The Y value equals 1. In step 3, move into the page 6 units; this is the negative Z direction, so the Z value equals -6. The coordinates for point B are $(-4, 1, -6)$.

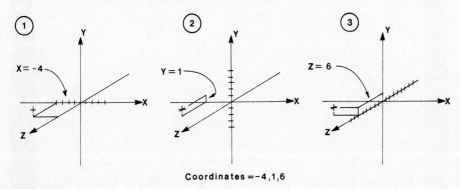

Coordinates $=-4,1,6$

FIGURE 2-21   How to locate a point on a three-dimensional XYZ axis system whose coordinates are $(-4, 1, -6)$.

2-9   ABSOLUTE AND RELATIVE AXIS SYSTEMS

The *absolute axis system* of a computer graphic system is the basic reference system built into the computer by the manufacturer. All calculations and transactions done by the computer are made in reference to the absolute axis system.

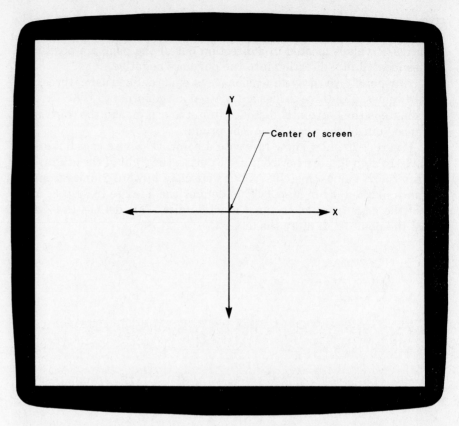

FIGURE 2-22 Absolute coordinate system.

FIGURE 2-23 Relative coordinate system.

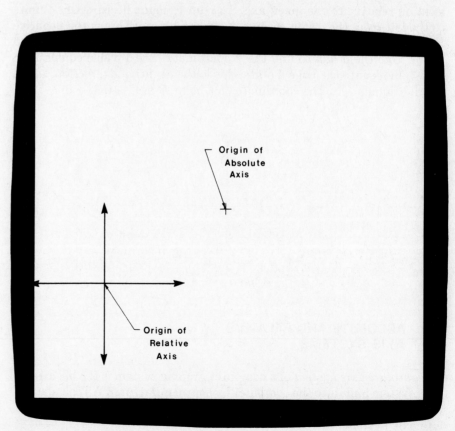

Most computer manufacturers set the origin location for the absolute system in the center of the screen. This is why, when a system is first turned on, the cursor, which represents the origin, appears in the center of the screen (Figure 2-22).

The absolute axis never moves but remains fixed in the predetermined location. Other axis systems, called *relative axis systems,* can be created by moving the original axis. In fact, the absolute axis system does not move, but disappears from the screen and the newly created relative system appears (Figure 2-23).

Relative axis systems are set up for the convenience of the user. In Figure 2-24 a relative axis system is located in the lower left corner of the screen. This means that the screen essentially shows only the first quadrant of the axis system and all input values for X and Y coordinates will be positive.

Some computer graphic systems allow several axis systems to exist on the same screen. This is helpful when working with multiple orthographic views and is discussed in Chapter 9.

FIGURE 2-24   Relative coordinate system.

Problems

For Problems P2-1 through P2-4, draw an XY axis system. Locate the origin in the center of the screen. Add the following points and, if possible, label each point with its letter and coordinate values. This exercise can be done using graph paper.

P2-1 All values are in inches.

(a) (2, 2)   (f) (4, −1)
(b) (−2, 2)   (g) (2, −3)
(c) (2, −2)   (h) (−1, −1)
(d) (−2, −2)   (i) (−1, 4)
(e) (1, 4)   (j) (−3, 1)

P2-2 All values are in millimeters.

(a) (50, 50)   (f) (100, −25)
(b) (−50, 50)   (g) (50, −75)
(c) (50, −50)   (h) (−25, −25)
(d) (−50, −50)   (i) (−25, 100)
(e) (25, 100)   (j) (−75, 25)

P2-3 All values are in inches.

(a) (1.50, 3.25)   (f) (3.75, −2.75)
(b) (0.50, −2.13)   (g) (−1.38, 3.50)
(c) (−2.25, −1.00)   (h) (−2.63, −1.75)
(d) (−1.75, 0.88)   (i) (0.13, 0.56)
(e) (3.13, 1.44)   (j) (−0.63, −0.44

P2-4 All values are in millimeters.

(a) (40, 82)   (f) (91, −62)
(b) (12, −65)   (g) (−37, 81)
(c) (−65, −25)   (h) (−72, −31)
(d) (−35, 20)   (i) (3, 5)
(e) (82, 33)   (j) (−13, −9)

For problems P2-5 through P2-8, list the points shown in Figures P2-5 through P2-8 along with their respective coordinates. For example, point *A* in Figure P2-5 would be listed as

1. *A* (2, 8)

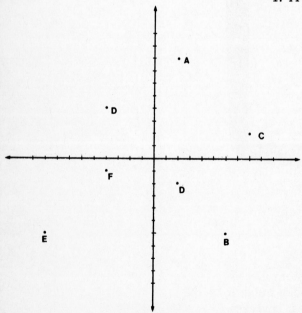

FIGURE P2-5

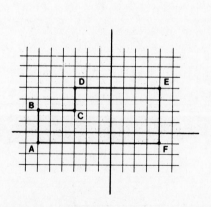

FIGURE P2-6

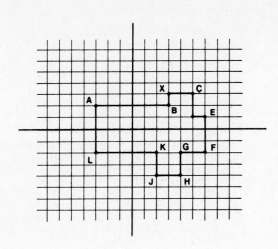

FIGURE P2-7

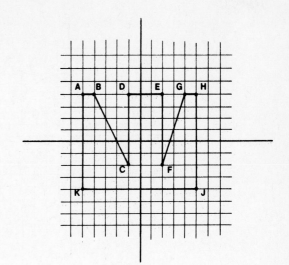

FIGURE P2-8

For Problems P2-9 through P2-12, draw an *XY* axis system. Locate the origin at the center of the screen. Add the following points and, if possible, label each point with its letter and coordinate values.

P2-9   All values are in inches and degrees (polar coordinates).

(a) (2, 45)	**(f)** (3, 100)
(b) (2, 135)	**(g)** (1, 150)
(c) (2, 225)	**(h)** (3, 200)
(d) (2, 315)	**(i)** (1, 250)
(e) (1, 30)	**(j)** (3, 330)

P2-10   All values are in millimeters and degrees (polar coordinates).

(a) (50, 45)	**(f)** (75, 100)
(b) (50, 135)	**(g)** (25, 150)
(c) (50, 225)	**(h)** (75, 200)
(d) (50, 315)	**(i)** (25, 250)
(e) (25, 30)	**(j)** (75, 330)

P2-11   All values are in inches and degrees (polar coordinates).

(a) (1.5, 30)	**(f)** (3.25, 327)
(b) (−1.0, 55)	**(g)** (−1.80, 15)
(c) (2.3, −40)	**(h)** (4.13, 227)
(d) (0.75, 160)	**(i)** (−0.30, 160)
(e) (−2.15, 200)	**(j)** (2.60, 290)

P2-12   All values are in millimeters and degrees (polar coordinates).

(a) (37, 30)	**(f)** (88, 327)
(b) (−25, 55)	**(g)** (−42, 15)
(c) (62, −40)	**(h)** (107, 227)
(d) (20, 160)	**(i)** (−8, 160)
(e) (−73, 200)	**(j)** (74, 290)

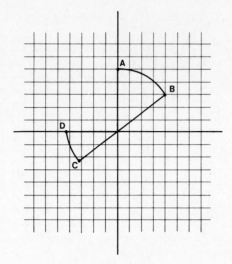

FIGURE P2-13   **FIGURE P2-14**

For Problems P2-13 and P2-14, list the points shown in Figures P2-13 and P2-14 along with their respective polar coordinates. Approximate the location of points not directly on a grid intersection. For example, point A in Figure P2-13 would be listed as

    1. *A* (5, 90)

P2-15   Figure P2-15 shows 10 points. List the points and their coordinate values relative to both axis systems: the absolute system and the system rotated 45° counterclockwise. For example, point *A* would be listed as

Point	Absolute	Rotated
1. *A*	(2, 5)	(5, 2)

FIGURE P2-15

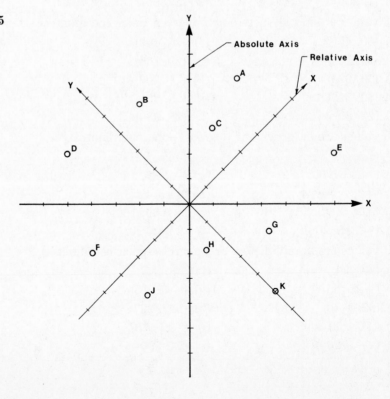

P2-16   Figure P2-16 shows 10 points. List the points and their coordinate values relative to both axis systems: the absolute system
and the relative system. For example, point A would be listed as

Point	Absolute	Relative
1. *A*	(1, 5)	(9, 11)

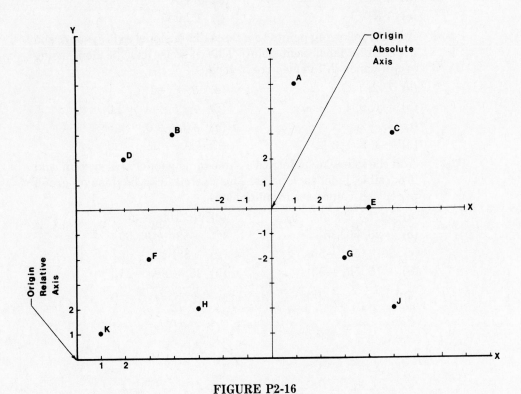

FIGURE P2-16

P2-17   The following points and coordinate values are based on an absolute coordinate system whose origin is at the center of the screen.
List the points with their coordinate values relative to the absolute coordinate system and with their coordinates relative to a
relative coordinate system whose origin is located at $(-7, -6)$
from the origin of the absolute system.

(a) (1, 1)   (f) $(-2, -3)$
(b) $(2, -1)$ (g) $(1, -5)$
(c) (3, 3)   (h) $(5, -2)$
(d) $(-2, 2)$ (i) $(-2, 1)$
(e) $(-5, -1)$ (j) $(-5, 2)$

P2-18   Express the following values in words and then as they would be
printed on a drawing.

(a) 0.003 ampere   (f) 0.000 000 000 025 farad
(b) 0.000 000 004 farad   (g) 4,000,000 ohms
(c) 1000 ohms   (h) 0.000 003 farad
(d) 10,000 watts   (i) 0.025 ampere
(e) 2,500,000 ohms   (j) 1500 volts

P2-19 Write the numerical equivalent of the following. Be sure to include the units of measure.

(a) 4 PF   (f) 1.5 Ω

(b) 3 KV   (g) 3 NF

(c) 1.5 mA   (h) 1.75 MΩ

(d) 2 KΩ   (i) 2 UA

(e) 6 K(W)   (j) 1 MW

P2-20 Add the following points to a three-dimensional axis system, and if possible, label each point. This exercise can be done using graph paper. All values are in inches.

(a) 2, 2, 1   (e) .5, 4, −1.5

(b) −1, 2, 1   (f) −2.5, −1.5, 2.0

(c) 3, −2, −2   (g) −3.5, 2.0, −2.5

(d) −3, 2, −2   (h) 1.0, −3.5, −1.0

P2-21 Add the following points to a three-dimensional axis system, and if possible, label each point. This exercise can be done on graph paper. All values are in millimeters.

(a) 50, 50, 25   (e) 12, 100, −36

(b) −25, 50, 25   (f) −62, −36, 50

(c) 75, −50, −50   (g) −88, 50, −62

(d) −75, 50, −50   (h) 25, −88, −25

3

Two-Dimensional Constructions

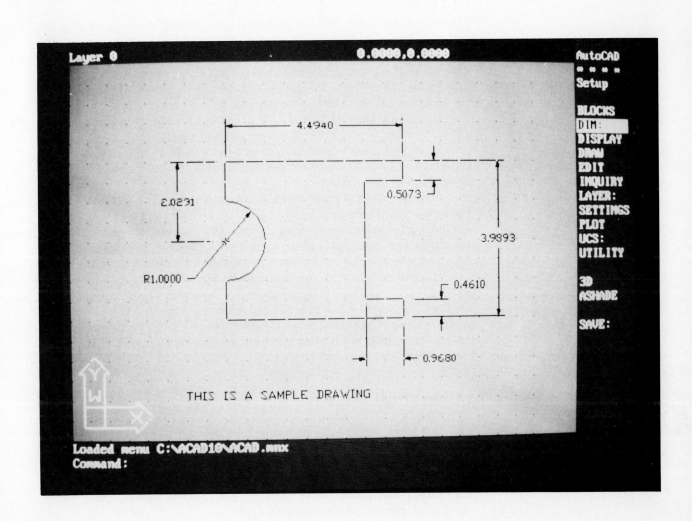

3-1 INTRODUCTION

In this chapter we explain how to draw basic two-dimensional shapes, including horizontal, vertical, and slanted lines, as well as, arcs, circles, and irregular shapes. We also explain how to combine these shapes to form planes.

3-2 HORIZONTAL LINES

There are three ways to draw horizontal lines using a computer graphic system: a line of infinite length through a point, a line from a known point with a specified length and direction, and a line from one known point to another known point. The input format for these three methods varies from system to system, and not all systems can utilize all three methods, but in general, these three methods apply.

Horizontal lines of infinite length may be drawn by specifying only a Y-axis value. A horizontal line has only a Y-axis value.

Figure 3-1 shows three horizontal lines: one at $Y = 3$, one at $Y = 0$, and one at $Y = -3$. These lines are of infinite length, so they run completely across the screen.

Figure 3-2 shows a horizontal line drawn between two known points. The input coordinators of the two points are $(-3, 3)$ and $(3, 3)$. The system would then be instructed to draw a line between these two points. Figure 3-3 shows four horizontal lines, each drawn between two given points. Note that the Y value is the same for both points on a horizontal line.

Horizontal lines can also be drawn by specifying the location of one point, a length, and a direction. Figure 3-4 illustrates the procedure. First a point of location $(-2.5, 2)$ is defined, then a length of 5 is entered, and finally, a direction to the right of the point is specified. Specifying a point, length, and direction gives an accurate line of known length that can be incorporated into a drawing.

3-3 VERTICAL LINES

There are three ways to draw vertical lines using a computer graphic system: a line of infinite length through a point, a line from one known point to another known point, and a line from a known point with a specified length and direction. The format for the required inputs varies with each individual system, and not all systems can operate with all three methods, but in general, these methods apply.

Vertical lines of infinite length have an X value but no Y value. The Y value is implied to be 0. Figure 3-5 shows three vertical lines: one at $X = 3$, one at $X = 0$, and one at $X = -3$. These lines are infinite, so they can run completely across the screen. Figure 3-6 shows a vertical

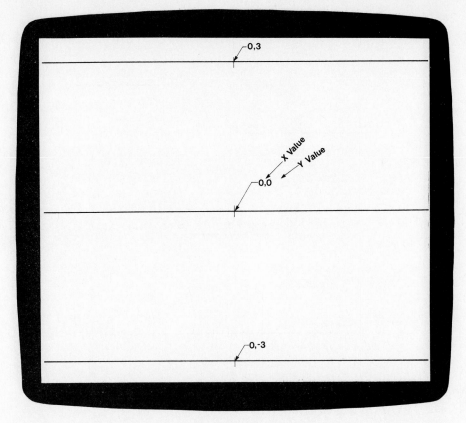

FIGURE 3-1 Horizontal lines of infinite length.

FIGURE 3-2 Horizontal line between two known points.

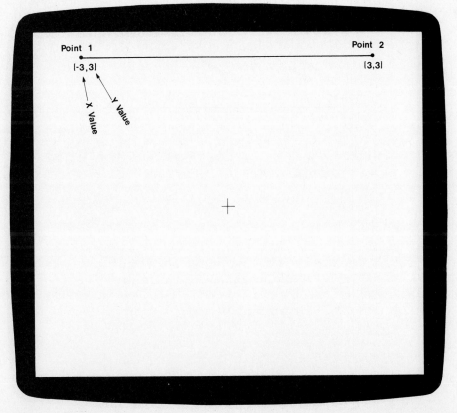

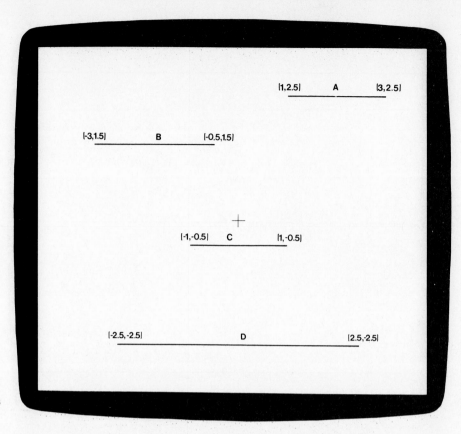

FIGURE 3-3 Horizontal lines between two known points.

FIGURE 3-4 Horizontal lines created by specifying a point, a length, and a direction.

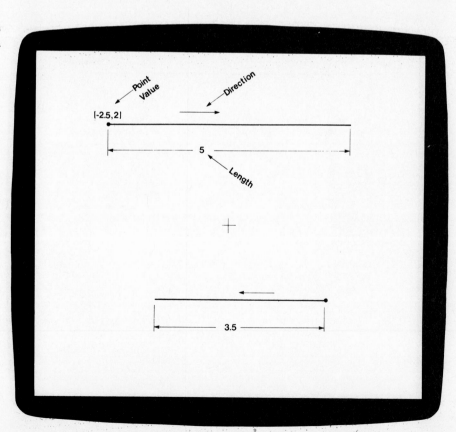

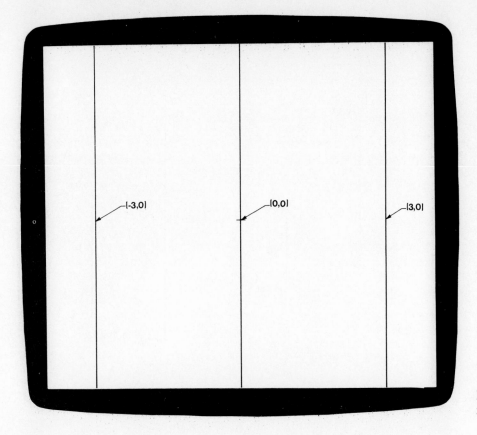

FIGURE 3-5 Vertical lines of infinite length.

FIGURE 3-6 Vertical line between two known points.

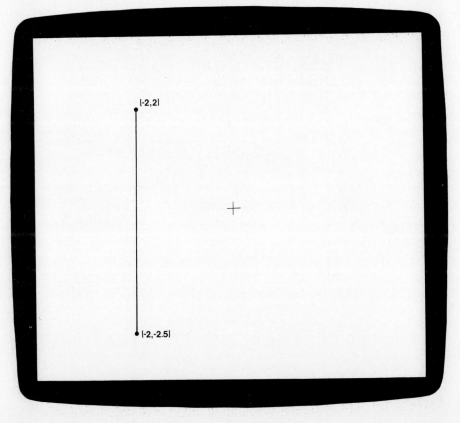

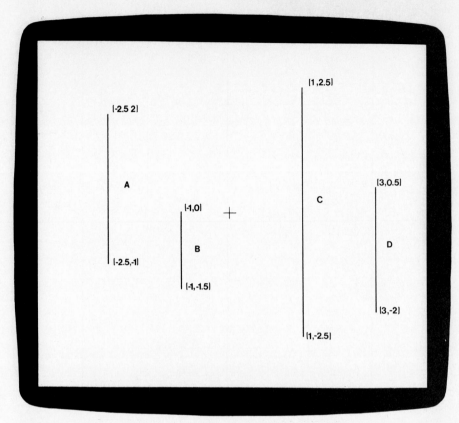

FIGURE 3-7 Vertical lines between two known points.

FIGURE 3-8 Vertical lines created by specifying a point, a length, and a direction.

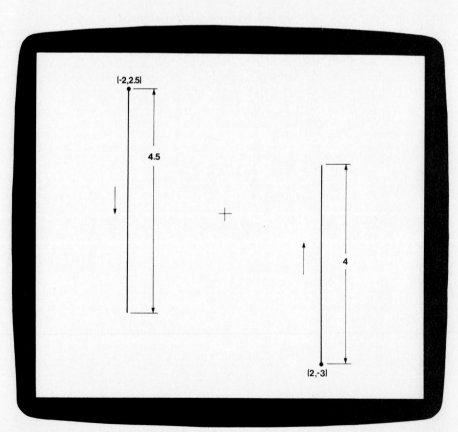

line between two points (2, 2) and (−2, −2.5). First the coordinate values for the points would be entered, then the system would be instructed to draw a line between the two points. Figure 3-7 shows four vertical lines. Each was drawn between two given points. Note that the X value is the same for all points on the given vertical line.

Vertical lines can also be drawn by specifying the location of one point, a length, and a direction. Figure 3-8 illustrates. First a point (2, −3) is defined, then a length 4, and finally a direction toward the top of the screen. It is also possible to direct the line toward the bottom of the screen.

3-4   COMBINING HORIZONTAL AND VERTICAL LINES

Horizontal and vertical lines may be combined to form shapes as shown in Figures 3-9 and 3-10. In Figure 3-9 the lines were drawn using lines between known points. Figure 3-10 was drawn giving each line a point, a length, and a direction.

Rectangular shapes can be created using lines of infinite length. Excess line length (the part of each line that extends beyond the needed length) can be erased or trimmed to create the desired final shape. The term "trim" refers to erasing part, but not all, of a drawn feature.

Figures 3-11 and 3-12 are more complex shapes that combine horizontal and vertical lines. Figure 3-11 was created using point-to-point lines and Figure 3-12 use lines specified using a point, length, and a direction.

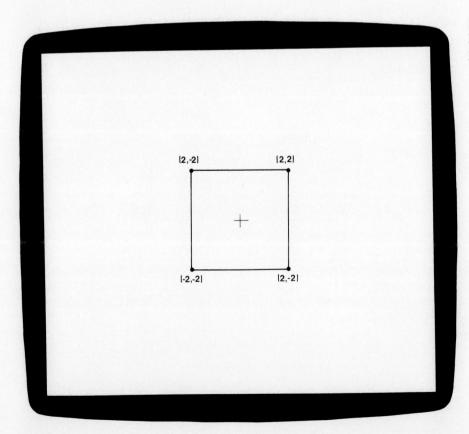

FIGURE 3-9 Shape formed by lines drawn between known points.

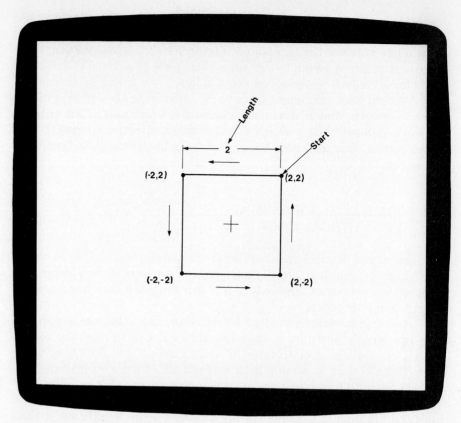

FIGURE 3-10 Shape formed by lines specified using a point, a length, and a direction.

FIGURE 3-11 Point-to-point lines.

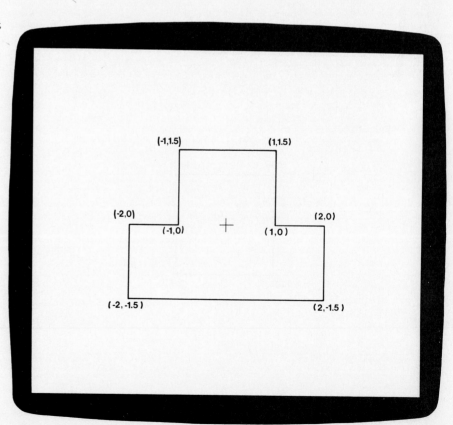

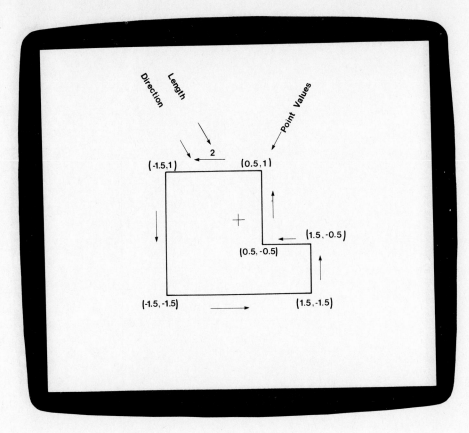

FIGURE 3-12 Point, length, and direction lines.

3-5 SLANTED LINES

Slanted lines can be drawn in one of two ways: point to point and by specifying a point location length and angle. Figure 3-13 shows two slanted lines drawn between known points. This method uses only linear inputs; that is, the points are defined in terms of the XY axis. No angular values are used.

Figure 3-14 shows a slanted line drawn by defining a point location, a length, and an angle. The angle was defined in degrees with the counter-clockwise direction as positive. (Angular values as used in computer graphic systems were discussed in Section 2-7.) Most technical drawings use angular and linear dimension rather than only linear dimensions. However, many numerically controlled machines require point-to-point type inputs, so both systems are important.

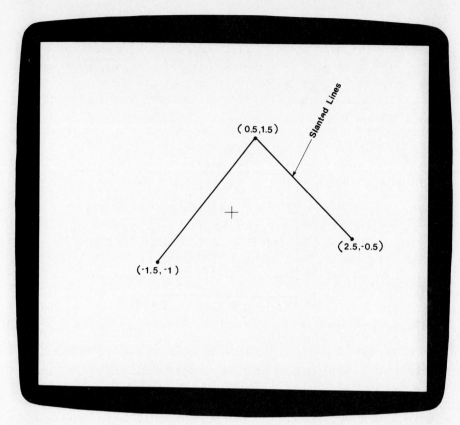

FIGURE 3-13 Slanted lines created by joining known points.

FIGURE 3-14 Slanted line created by defining a point, line length, and angle.

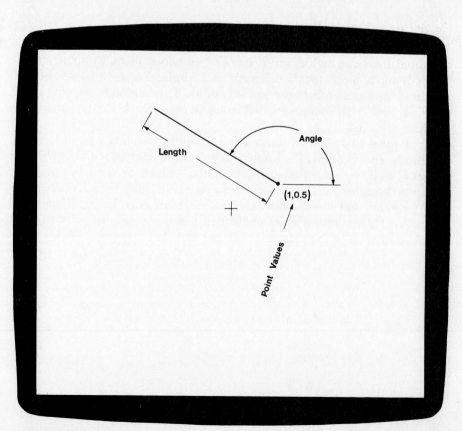

3-6   CIRCLES

Circles are drawn by specifying a centerpoint location and a size. The location is given by defining a point location and the size by defining a radius or diameter depending on the system. In Figure 3-15 a circle was drawn around the screen centerpoint (0, 0) using a radius value of 0.75.

Circles drawn by computer graphic systems are not pure mathematical shapes. (They are not exactly equal to the loci of all points about a centerpoint at a constant radius.) They are a series of short straight lines. The more straight lines, the smoother the circular shape. This is referred to as a screen resolution. The higher the resolution, the smoother the circle.

Systems that draw circles as a series of straight lines require a series of inputs specifying a distance and an angle. The procedure is to create a series of interconnected lines that appear as a circle. For example, in Figure 3-16, a distance of 10 screen units and an angle of 15° were used to create a circle. Figure 3-17 shows a photograph of a circle drawn this way using an Apple IIe computer working in LOGO.

The total value of the angles used to draw the circle shown in Figure 3-16 must equal 360°. Therefore, the distance and 15° angle must be repeated 24 times to form a complete circle. The diameter size of the circle is controlled by the distance input.

Figure 3-18(a) shows three circles. Each has a different centerpoint location and size. The input for each was an X value, a Y value, and a radius value. Figure 3-18(b) shows a circle drawn on a very high-resolution screen (Tektronix-618 using an IBM 8084 system working with CADAM).

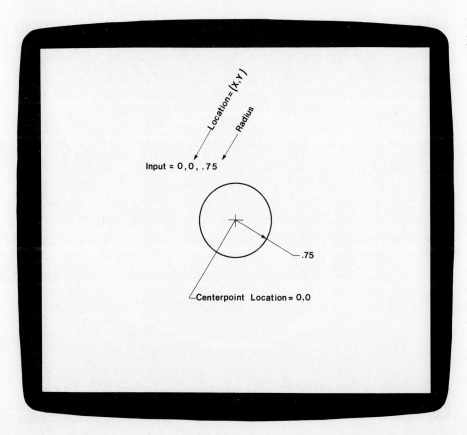

FIGURE 3-15 Circle with inputs for location and radius.

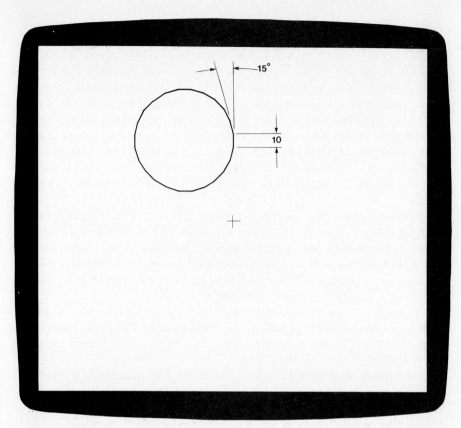

FIGURE 3-16 Circle formed by short straight-line segments.

FIGURE 3-17 Circle formed using short straight-line circles (Apple IIe computer).

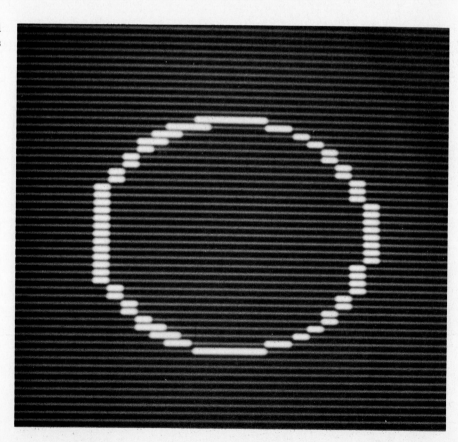

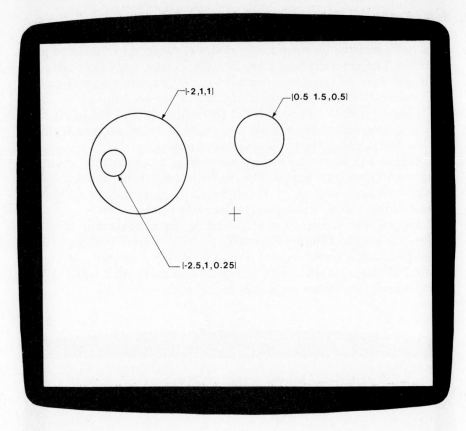

FIGURE 3-18(a)   Three different circles with their inputs.

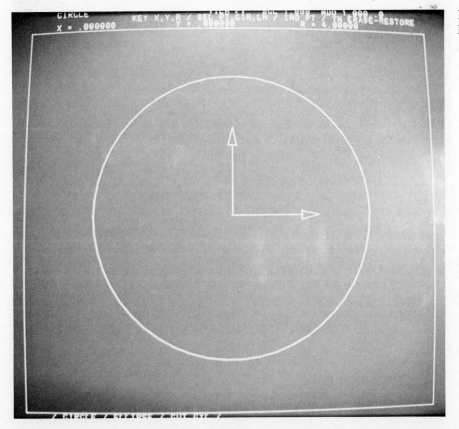

FIGURE 3-18(b)   Circle on a high-resolution screen.

3-7 ARCS

Arcs, as interpreted by computer graphics systems, are considered parts of circles. The inputs required are location, radius, and arc length. The arc length may be specified in degrees or by first drawing a circle and then erasing that portion of the circle that is not needed.

Erasing part of a circle is called *trimming.* To trim a circle, it must first be divided into segments. The arc would be considered one segment, and the remainder of the circle a second segment.

Figure 3-19 shows two arcs: one defined using degrees, the other using starting and end points. The arc centerpoint is located using *X, Y* coordinates. Each also uses a radius input. In general, arcs are defined using a radius value, circles using a diameter value.

Figure 3-20 shows an arc created by interconnecting lines and angles. The length of the arc is controlled by limiting the number of lines. A complete circle would require a total angle input of 360°. The arc in Figure 3-20 has a length of 90°, so the total angular must equal 90°. In this example, this means six angles of 15° each.

FIGURE 3-19 Two different methods for drawing an arc.

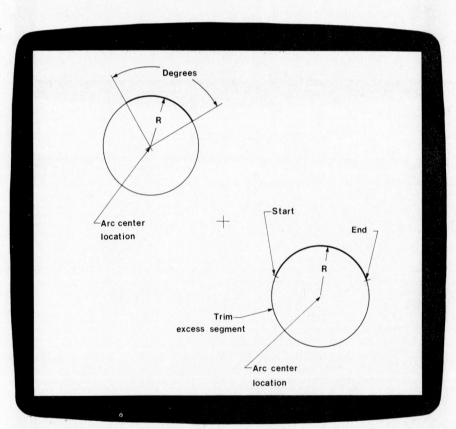

3-8 FILLETS AND ROUNDS (CORNERS)

A *fillet* is a concave arc and a *round* is a convex arc (Figure 3-21). They are often referred to as *corners,* particularly if they are 90° arcs. Many computer systems have a separate function specifically to create fillets and rounds.

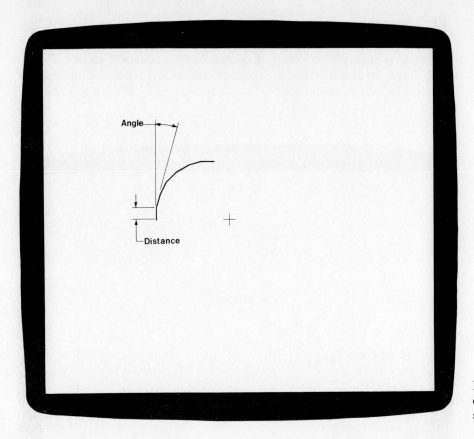

FIGURE 3-20 A 90° arc created by short straight-line segments.

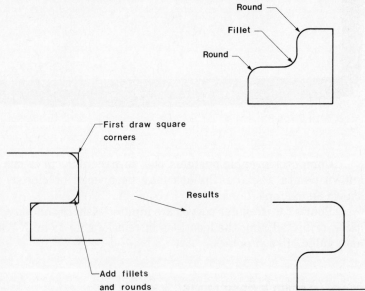

FIGURE 3-21 Fillets and rounds.

Fillets and rounds are drawn by adding arcs between straight-line corners and then erasing the square corner beyond the arc. Figure 3-21 shows how this process works. The shape is first drawn square, that is, as if it were constructed entirely of straight lines; then the corners are added. Some systems will automatically erase the excess straight lines, whereas others require line trimming. Centerlines through the arc center-point can be used to help define the beginning and end of the arc for trimming purposes.

3-9 IRREGULAR CURVES

Irregular curves are curves that do not have a constant radius. Figure 3-22 shows an irregular curve. Irregular curves are defined using a series of point locations, then constructing a smooth line between the points.

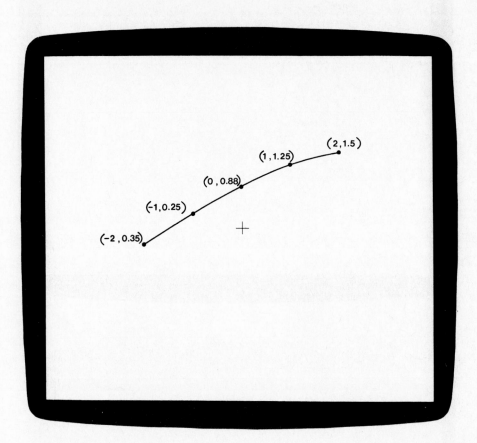

FIGURE 3-22 Irregular curve.

    Computer graphic systems call an irregular curve drawn between known points a *spline*. Splines may be opened or closed. Figure 3-23 shows an example of each.

    Points for irregular curves are inputed either randomly, using a digitizer, or by defining the locations in relationship to an XY axis (coordinate value of each point).

3-10 SAMPLE PROBLEMS

Figure 3-24 shows a shape that is made up of several different shapes: horizontal lines, vertical lines, and a circle. Shape 1 was drawn using point inputs, and shape 2 was drawn using point location, length, and distance inputs.

    Figure 3-25 is a shape drawn by combining horizontal lines, vertical lines, an arc, a circle, and two corners (fillets). It is typical of shapes used in technical drawing.

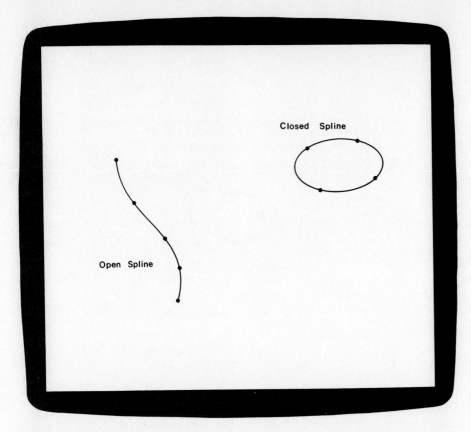

FIGURE 3-23   Open and closed splines.

FIGURE 3-24 Two shapes made by combining lines and circles.

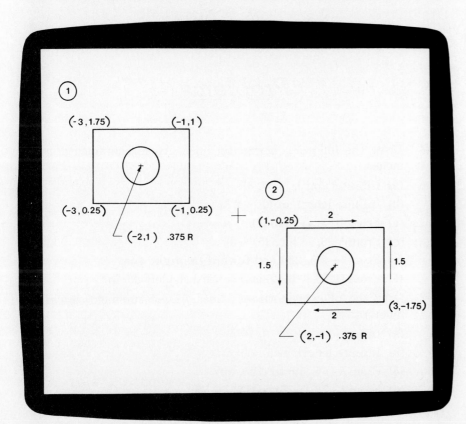

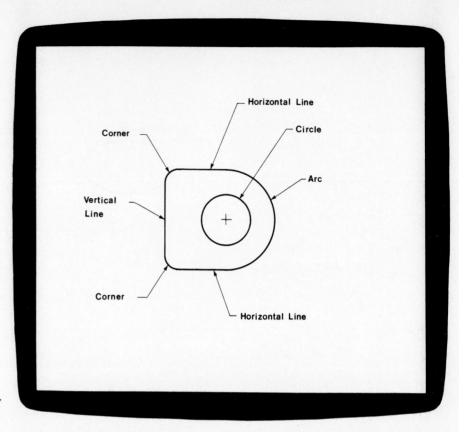

FIGURE 3-25 Shape made by combining other basic shapes.

Problems

P3-1 Draw the following horizontal lines. All measurements are in inches.

(a) Infinite length at (0, 4)

(b) Infinite length at (0, −2.5)

(c) From (−3, 2) to (5, 2)

(d) From (−3, −3) to (5, −3)

(e) From (−2.5, 2.5) and toward the right 4 in.

(f) From (1.75, −0.50) and toward the left 2.5 in.

P3-2 Draw the following horizontal lines. All measurements are in millimeters.

(a) Infinite length at (0, 100)

(b) Infinite length at (0, 60)

(c) From (−75, 50) to (125, 50)

(d) From (−75, −75) to (125, −75)

(e) From (−60, 60) and toward the righrt 110 mm

(f) From (45, −13) and toward the left 75 mm

P3-3   Draw the following vertical lines. All measurements are in inches.

(a) Infinite length at (3, 0)

(b) Infinite length at (−2, 0)

(c) From (−2.5, 2) to (−2.5, −2)

(d) From (2, 1.5) to (2, −2.5)

(e) From (−0.5, 2) and toward the bottom 4 in.

(f) From (1, 1.3) and upward 3 in.

P3-4   Draw the following vertical lines. All measurements are in millimeters.

(a) Infinite length at (75, 0)

(b) Infinite length at (−50, 0)

(c) From (−60, 50) to (−60, −50)

(d) From (50, 38) to (50, −63)

(e) From (−13, 50) and toward the bottom 105 mm

(f) From (25, 33) and upward 75 mm

P3-5   Draw the following slanted lines. All measurements are in inches.

(a) From (−2, 1) to (0, 2)

(b) From (−2.5, −2.5) to (2.5, 0.5)

(c) From (0.5, 3) to (0, 0)

(d) From (−1.5, −0.5) to the right 3 in. at 15° below the horizontal

(e) From (0.5, −2.0) upward 4 in. at 20° to the right of vertical

P3-6   Draw the following slanted lines. All measurements are in millimeters.

(a) From (−50, 25) to (0, 50)

(b) From (−62, −62) to (62, 13)

(c) From (13, 75) to (0, 0)

(d) From (−33, −13) to the right 75 mm at 15° below the horizontal

(e) From (13, −50) upward 100 mm at 20° to the right of the vertical

P3-7   Draw the following circles. All measurements are in inches.

(a) At (0, 0) with radius = 1.00

(b) At (−1, 0) with diameter = 4.00

(c) At (2, 2) with radius = 0.50

(d) At (2, −0.5) with diameter = 0.50

(e) At (1.5, 1) with radius = 1.25

(f) At (1, −2) with diameter = 1.50

P3-8   Draw the following circles. All measurements are in millimeters.

(a) At (0, 0) with radius = 25

(b) At (−25, 0) with diameter = 100

(c) At (50, 50) with radius = 13

(d) At (50, −13) with diameter = 13

(e) At (38, 25) with radius = 33

(f) At (25, −50) with diameter = 40

P3-9   Draw the following irregular curves. Each curve is defined using
five points. All measurements are in inches.

(a) Curve 1

Point	Coordinates
1	(−3.0)
2	(−2, 1.2)
3	(−0.5, 2.3)
4	(0.7, 2.7)
5	(2, 3)

(b) Curve 2

Point	Coordinates
1	(−3, 0)
2	(−1, 1)
3	(0.3, 1.4)
4	(1.5, 1.3)
5	(2.7, 0)

(c) Curve 3

Point	Coordinates
1	(−2, 2.5)
2	(−1.3, 2.6)
3	(−0.5, 0.6)
4	(0, 0)
5	(0.6, −1)

(d) Curve 4

Point	Coordinates
1	(−1.2, −2)
2	(−1, −1.3)
3	(−0.6, −0.5)
4	(0, 0)
5	(1, 0.4)

P3-10   Draw the following irregular curves. Each curve is defined using
five points. All measurements are in millimeters.

(a) Curve 1

Point	Coordinates
1	(−75, 0)
2	(−50, 30)
3	(−13, 58)
4	(18, 68)
5	(50, 75)

(b) Curve 2

Point	Coordinates
1	(−75, 0)
2	(−25, 25)
3	(8, 36)
4	(38, 32)
5	(68, 0)

(c) Curve 3

Point	Coordinates
1	(−50, 62)
2	(−33, 66)
3	(−13, 16)
4	(0, 0)
5	16, −25)

(d) Curve 4

Point	Coordinates
1	(−30, −50)
2	(−25, −33)
3	(−16, −13)
4	(0, 0)
5	(25, 10)

P3-11   Draw Figure P3-11. All coordinate values and dimensions are in inches.

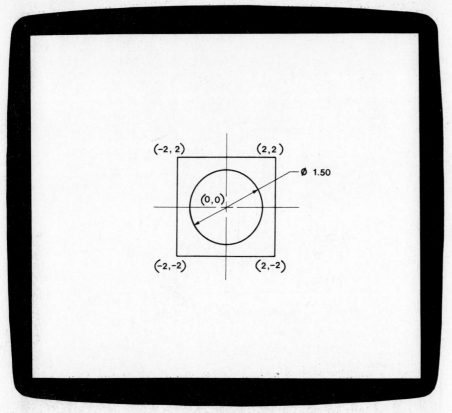

FIGURE P3-11

P3-12   Draw Figure P3-12. All coordinate values are in millimeters.

FIGURE P3-12

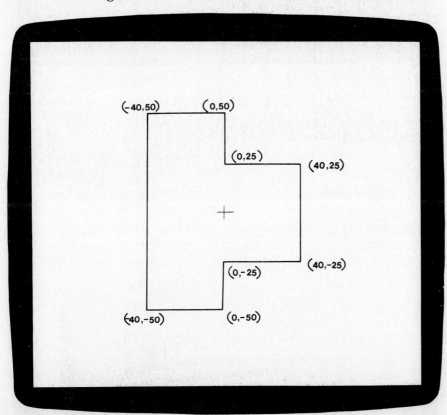

P3-13   Draw Figure P3-13. All coordinate values and dimensions are in millimeters.

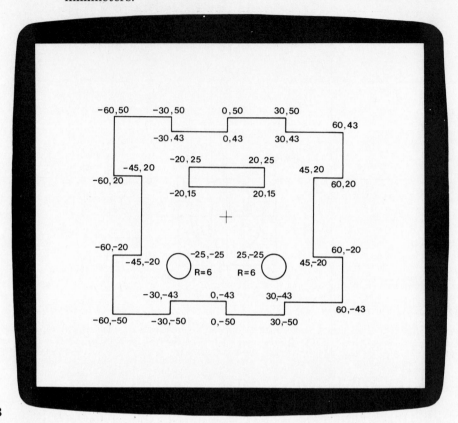

FIGURE P3-13

P3-14   Draw Figure P3-14. All coordinate values are in millimeters.

FIGURE P3-14

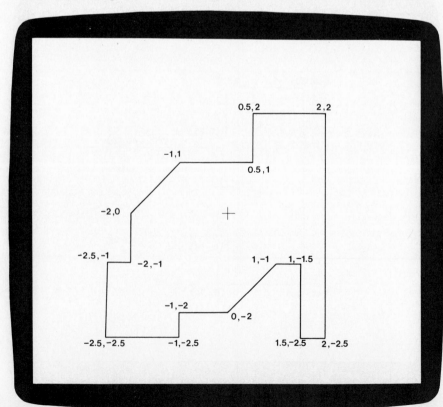

P3-15 Draw Figure P3-15. All dimensions are in millimeters. Each corner of the figure is identified by a number. The coordinate values for each numbered corner are listed below.

Point	Coordinates	Point	Coordinates
1	(−75, 60)	29	(30, −40)
2	(−60, 60)	30	(40, −25)
3	(−45, 53)	31	(12, −25)
4	(−45, 35)	32	(12, −43)
5	(−25, 35)	33	(−12, −43)
6	(−35, 17)	34	(−12, −25)
7	(−15, 17)	35	(−40, −25)
8	(−15, 47)	36	(−30, −40)
9	(15, 47)	37	(−30, −50)
10	(15, 17)	38	(−40, −50)
11	(35, 17)	39	(−60, −60)
12	(25, 35)	40	(−75, −60)
13	(45, 35)	41	(−75, −50)
14	(45, 53)	42	(−65, −50)
15	(60, 60)	43	(−65, −30)
16	(75, 60)	44	(−52, 30)
17	(75, 50)	45	(−52, 30)
18	(65, 50)	46	(−65, 30)
19	(65, 30)	47	(−65, 50)
20	(52, 30)	48	(−75, 50)
21	(52, −30)	49	(−40, 9)
22	(65, −30)	50	(−15, 9)
23	(65, −50)	51	(−15, 0)
24	(75, −50)	52	(15, 0)
25	(75, −60)	53	(15, 9)
26	(60, −60)	54	(40, 9)
27	(40, −50)	55	(40, −15)
28	(30, −50)	56	(−40, −15)

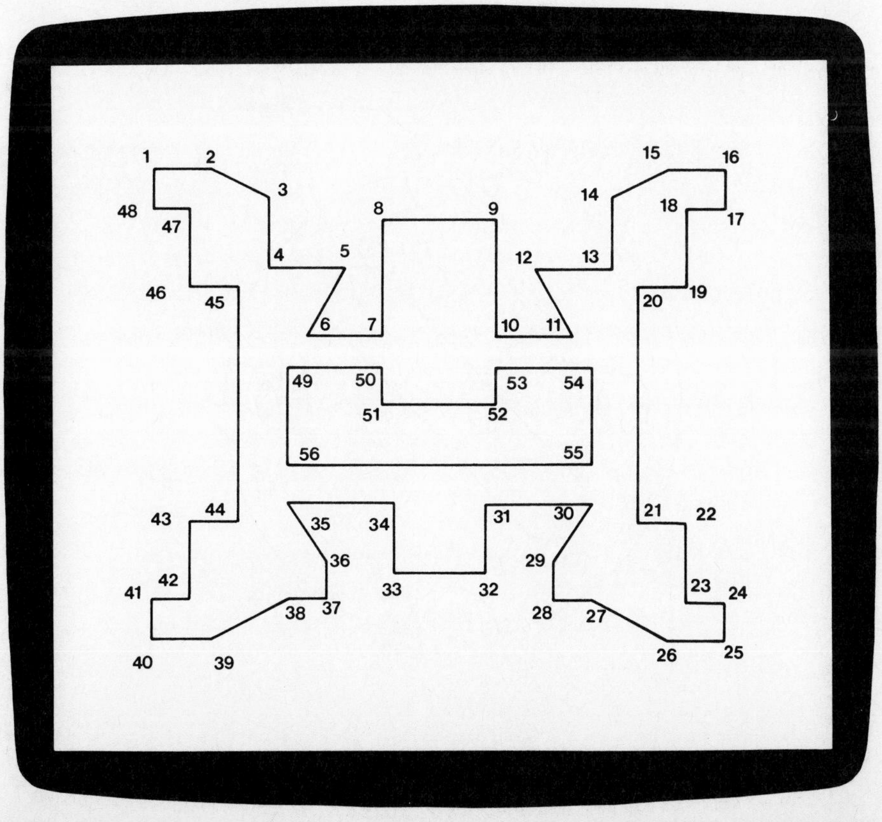

FIGURE P3-15

For Problems P3-16 through P3-19, draw Figures P3-16 through P3-19 based on the coordinate and dimension information given. The dimensional units are as noted below.

P3-16   Millimeters   P3-18   Inches
P3-17   Millimeters   P3-19   Millimeters

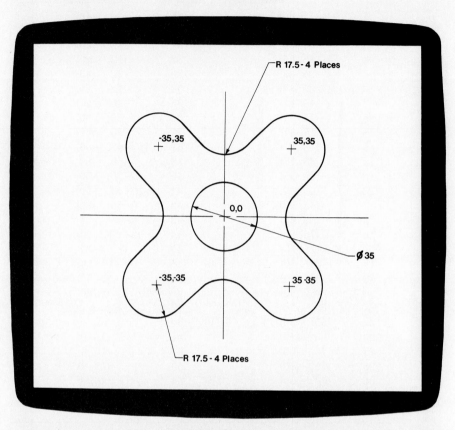

FIGURE P3-16

FIGURE P3-17

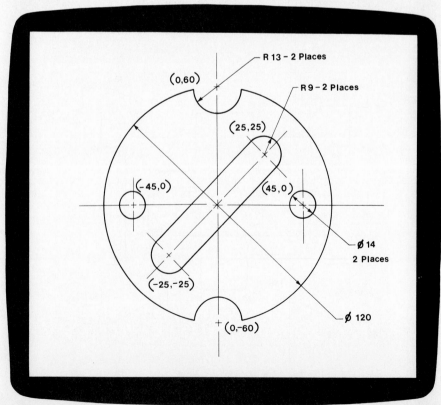

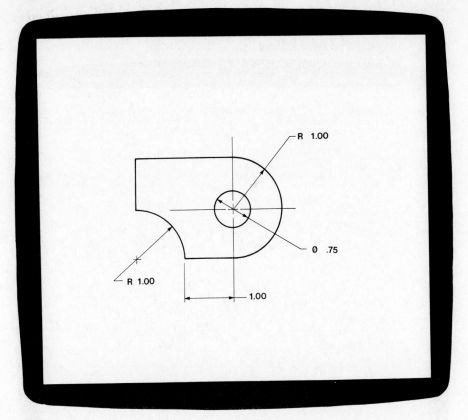

FIGURE P3-18

FIGURE P3-19

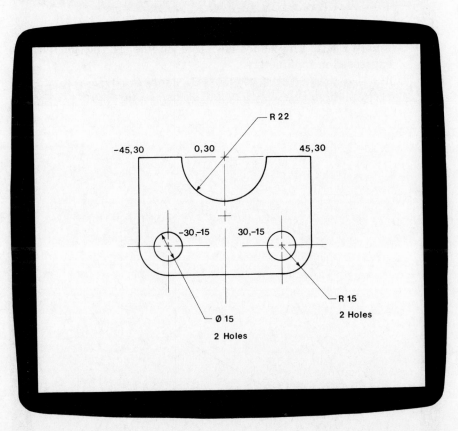

P3-20   Start with the defined *A* point location (−2, −1.50) and construct Figure P3-20. The values next to each line are the line's length expressed in inches.

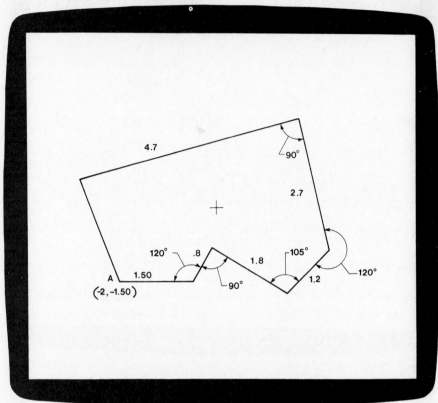

FIGURE P3-20

P3-21   Start with the defined *A* point location (−65, −60) and construct Figure P3-21. The values next to each line are the line's length expressed in millimeters.

FIGURE P3-21

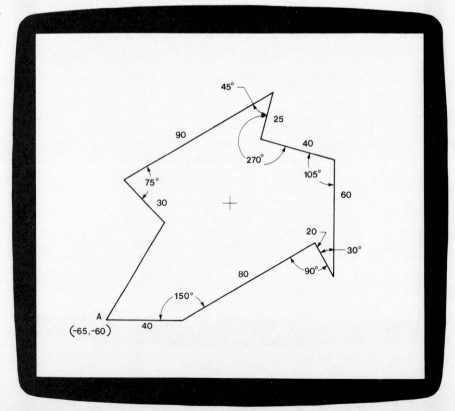

P3-22   Draw Figure P3-22. All coordinate and dimensional values are in inches.

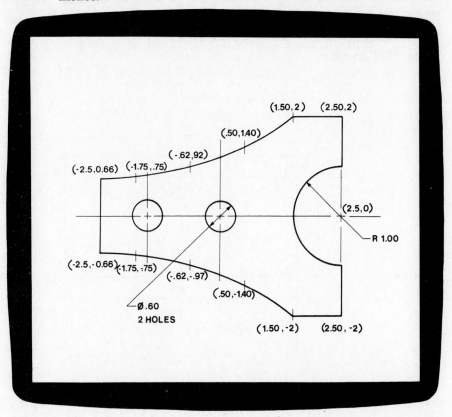

FIGURE P3-22

P3-23   Draw the CAM shown in Figure P3-23 based on the information given. Each point on the outer surface is defined using polar coordinates. All radius dimensions are in millimeters.

FIGURE P3-23

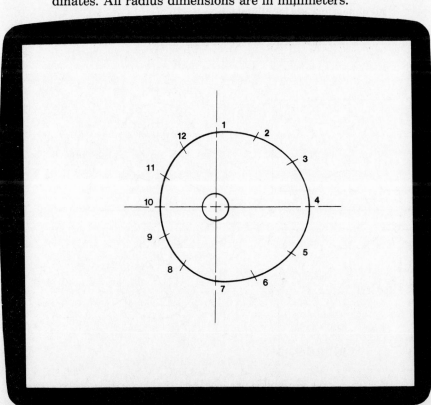

For problems P3-24 through P3-28, redraw Figures P3-24 through P3-28 using the dimensions given. The dimensional units are as noted below.

P3-24	Inches	P3-27	Millimeters
P3-25	Millimeters	P3-28	Millimeters
P3-26	Millimeters		

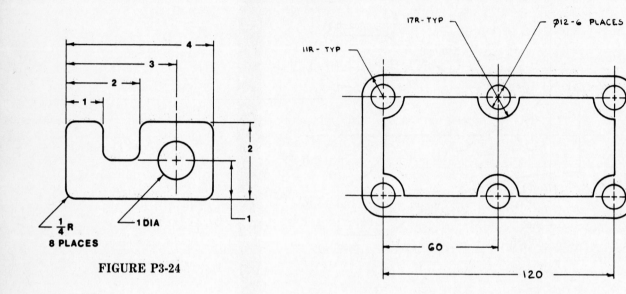

FIGURE P3-24

All dimensions are in millimeters.

FIGURE P3-25

All dimensions are in millimeters.

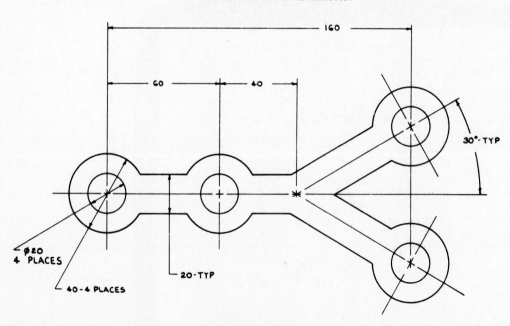

FIGURE P3-26

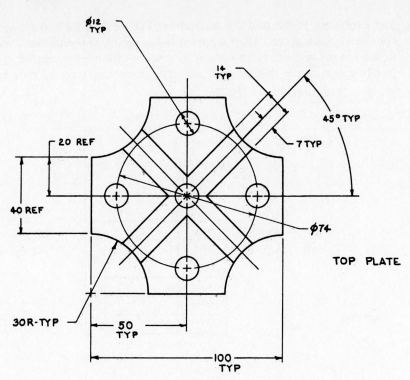

TOP PLATE

FIGURE P3-27

FIGURE P3-28

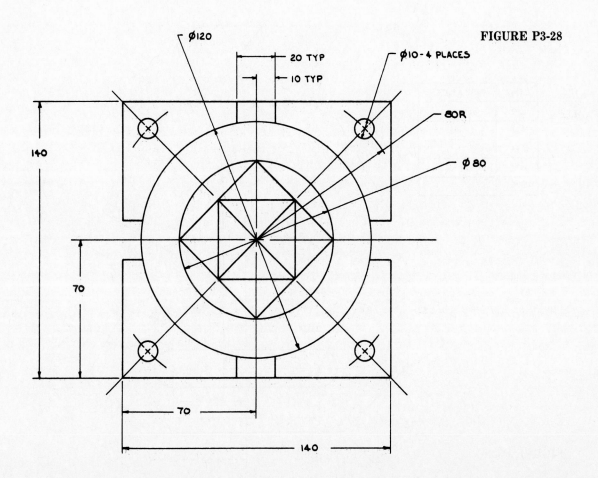

For Problems P3-29 and P3-30, redraw Figures P3-29 and P3-30 using the dimensions given. After completion, measure the distances specified in the charts and list their values. Use the system's measuring capabilities to measure the distances. The dimensional units are as noted below.

P3-29   Millimeters   **P3-30**   Inches

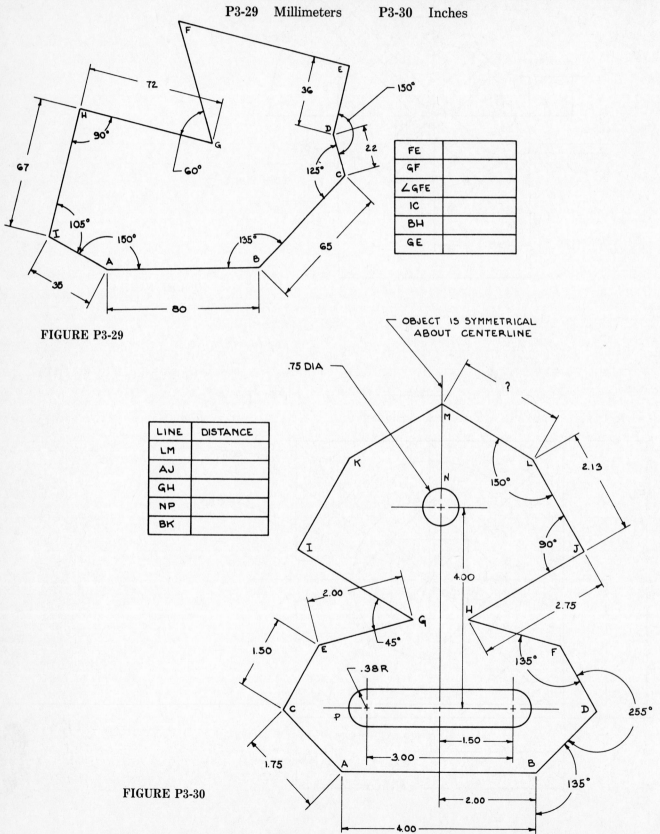

FE	
GF	
∠GFE	
IC	
BH	
GE	

FIGURE P3-29

LINE	DISTANCE
LM	
AJ	
GH	
NP	
BK	

FIGURE P3-30

4

Electronic Symbols

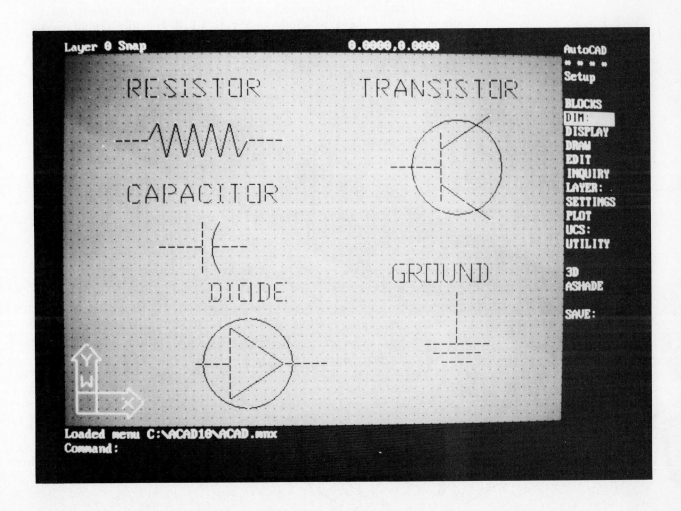

4-1 INTRODUCTION

Electronic symbols are two-dimensional representations of electronic components. They are used to help create drawings and schematics of electronic circuits.

4-2 SYMBOL LIBRARIES

Symbols are stored in a file called a *symbol library*. Many electronic CAD systems come with a symbol library built into their software. If no library is available, it is suggested that one be created by drawing each component referenced in this chapter, and filing them for future reference.

Electronic libraries are accessed in one of three ways, depending on the CAD system being used. Some systems present the symbols pictorially on a menu board, as shown in Figure 4-1. A specific symbol can be

FIGURE 4-1 Typical computer graphics setup.

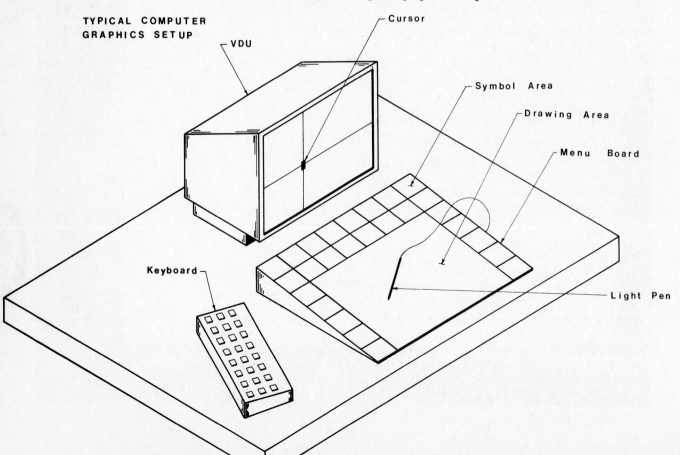

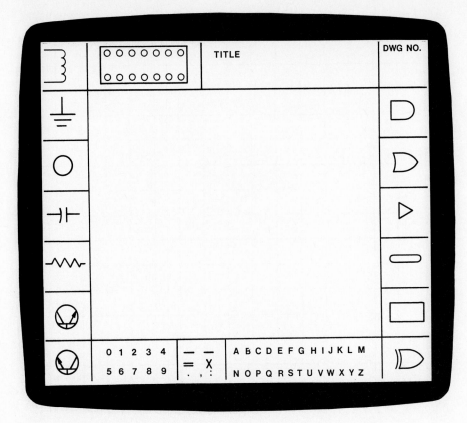

FIGURE 4-2 An example of symbols presented using a screen reference setup.

accessed by identifying the symbol using a light pen or mouse, and then pulling the symbol into the working area of the board. The symbol should then appear on the VDU screen.

A second method used to file and access electronic symbols is a screen reference system, as shown in Figure 4-2. This type of setup presents the symbols around the edge of the VDU screen surrounding the working area. The symbol is identified using a light pen or mouse and then moved into the working area.

A third method is to file each symbol by name and then access the file using the keyboard. For example, a resistor may be filed under RES, a capacitor CAP, and a transistor TRAN-NPN or TRAN-PNP, depending on the type.

4-3   GRID BACKGROUNDS

Most CAD systems have the ability to add a grid background to the screen. The grid patterns may be displayed in one of three patterns: perpendicular lines (Figure 4-3), edge line indications (Figure 4-4), and dots or points (Figure 4-5).

Grid backgrounds serve to help locate the symbols in easy-to-reference locations. All symbols will have an origin or reference point that can be used to locate the symbol within the work area. The symbol origins can be aligned with the intersections of the grid background to help assure even spacing, giving the drawing a neat, organized appearance. Figure 4-6 shows an IC chip and a transistor located on a dot background.

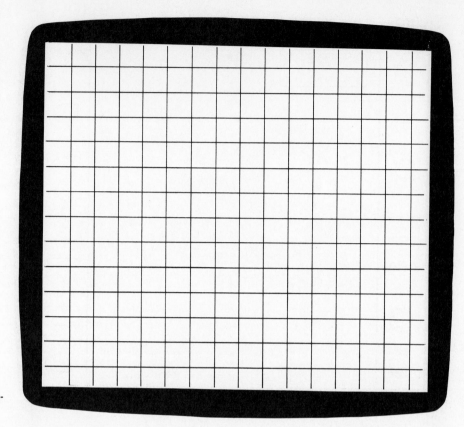

FIGURE 4-3 A line grid pattern.

FIGURE 4-4 An edge line grid pattern.

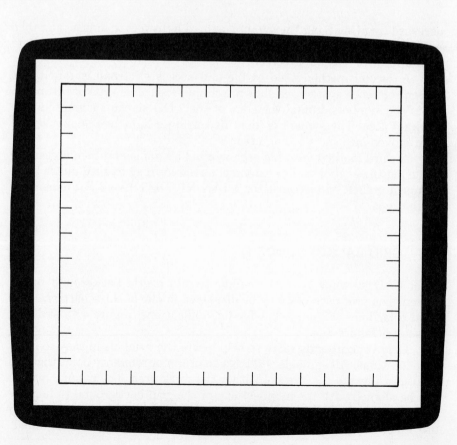

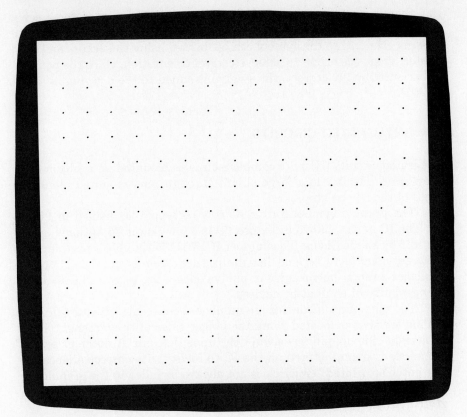

FIGURE 4-5 A dot grid pattern.

FIGURE 4-6 Symbols located using a dot grid.

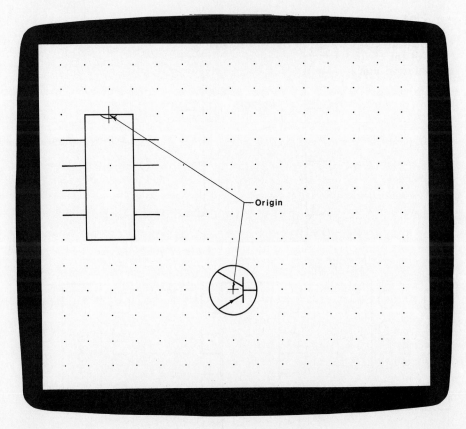

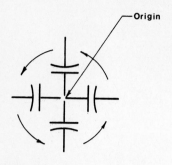

FIGURE 4-7 A capacitor symbol rotated about its symbol origin.

Symbols can be rotated about their origins as needed for a schematic diagram. CAD systems treat an individual symbol as a complete drawing element, so the symbol can be located using the origin, and then rotated about the origin to achieve correct orientation. Once the position and orientation are correct, the symbol is added to the drawing (see Figure 4-7).

4-4 INTEGRATED CIRCUITS

Integrated circuits (ICs) are complete circuits mounted on a silicon chip. They are connected to a large circuit through pins mounted along the edges of the silicon chip.

The graphic symbol for an IC is a rectangle, as shown in Figure 4-8. The IC device shown in Figure 4-8 is a low-power operational amplifier sold by Radio Shack. The number ICL7611 is the chip's part number. Different manufacturers and distributors use different numbers, but each publishes a replacement guide or equivalent listing to permit easy interchangeability of equivalent devices.

The pins of an IC device are numbered consecutively starting with 1. Pins are always located along the longer sides of the rectangle. If the IC is right-side up, pin 1 is always the upper left pin. A notch or printed semicircle is manufactured on the IC to identify its correct orientation. The notch or printed semicircle is not always included in the graphic representation of the IC. Figure 4-9 shows other possible symbols for intergrated circuits.

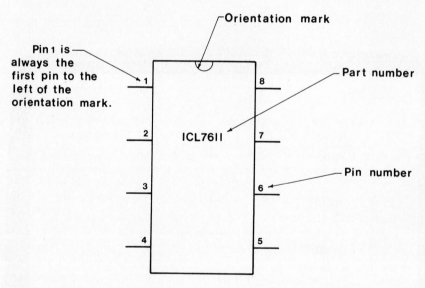

FIGURE 4-8 Graphic symbol for an integrated circuit.

FIGURE 4-9 Symbols for integrated circuits.

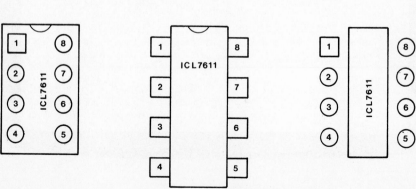

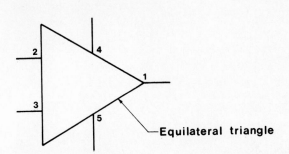

FIGURE 4-10 Graphic symbol for an amp-pere.

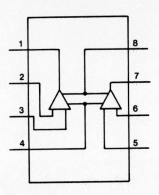

FIGURE 4-11 IC symbol that includes ampere functions.

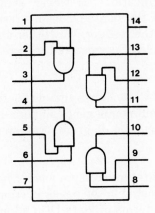

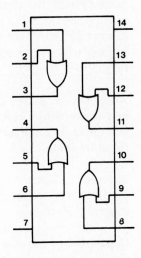

FIGURE 4-12 IC with logic symbols.

An amplifier (amp) is a type of IC device which is used so often that it has its own symbol, an equilateral triangle. Figure 4-10 shows an amp symbol. Remember that an equilateral triangle has three 60° angles and three equal-length sides. Pin numbers or amps vary as to specific functions.

Amps are often used individually or in groups as part of an IC. The actual chip is still rectangular in shape, and is represented symbolically by a rectangle, but the rectangle also includes the amp symbols. Figure 4-11 shows the symbol for such an IC.

Logic symbols (see Chapter 7) are often included within the rectangular IC symbol, to help define the IC's function. Figure 4-12 shows two examples of IC symbols that include logic symbols.

4-5 SYMBOL SHAPES

Most companies use symbol shapes that conform to the standards set by the American National Standards Institute (ANSI), 1430 Broadway, New York, NY 10018, or by Military Standards and Specifications, Naval Publications and Forms Center, 5801 Tabor Avenue, Philadelphia, PA 19120. The symbols presented in this book conform to both standards.

per Standards

Acceptable   Wrong

FIGURE 4-13 Acceptable and incorrect ways to draw electronic symbols.

There is no specific size for electronic symbols. They may be varied according to the drawing specifications. However, symbol proportions may *not* be varied. For example, the slanted lines of a resistor symbol are always 60° to the centerline of the resistor. Any other angle is incorrect. Figure 4-13 shows some example of correct and incorrect resistor symbols.

Antennas   Figure 4-14 shows the standard symbols for antennas. Antennas are generally considered input components and are usually placed in the upper left portion of a schematic diagram. Antennas are generally rated in ohms.

Batteries   Figure 4-15 shows the standard symbols for batteries. Batteries are power sources and are located in the power portion of a schematic diagram. Batteries are rated in terms of volts DC and amperes.

Capacitors   Figure 4-16 shows the standard symbols for capacitors. Capacitors are used to control current within a circuit and are rated in farads. Most schematics call for very small unit values. For example, 1 picofarad is equal to 0.000 000 000 001 farad. See Section 2-2 for an explanation of small-value prefixes.

FIGURE 4-14   Antenna symbols.

FIGURE 4-15   Battery symbols.

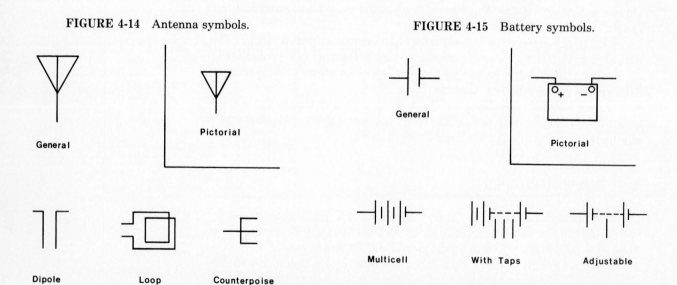

General

Pictorial

Dipole   Loop   Counterpoise

General

Pictorial

Multicell   With Taps   Adjustable

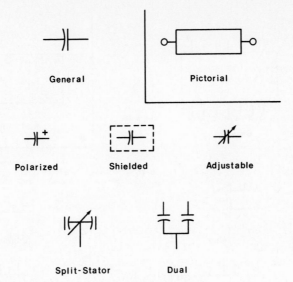

General   Pictorial

Polarized   Shielded   Adjustable

Split-Stator   Dual

FIGURE 4-16   Capacitor symbols.

Diodes   Figure 4-17 shows the standard symbols for diodes. Diodes are used to control current within a circuit and are rated in volts.

Grounds   Figure 4-18 shows the standard symbols for grounds. Grounds do not have unit values. They are generally located along the bottom of a schematic diagram.

Inductors   Figure 4-19 shows the standard symbols for inductors. Inductors are used to control current within a circuit and are rated in henrys.

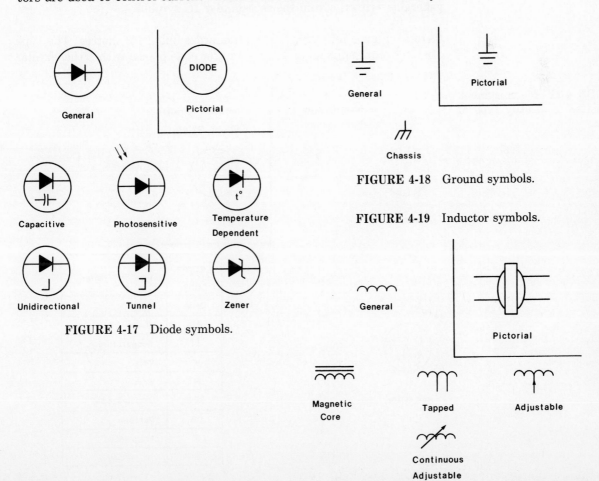

General   Pictorial

Capacitive   Photosensitive   Temperature
                                Dependent

Unidirectional   Tunnel   Zener

FIGURE 4-17   Diode symbols.

General   Pictorial

Chassis

FIGURE 4-18   Ground symbols.

FIGURE 4-19   Inductor symbols.

General   Pictorial

Magnetic   Tapped   Adjustable
Core

Continuous
Adjustable

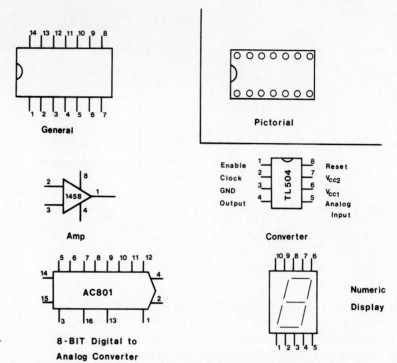

FIGURE 4-20 Intigrated circuit symbols.

Integrated circuits Figure 4-20 shows the standard symbols for integrated circuits. Integrated circuits (ICs) are identified by manufacturer's part number: for example, 80386 or HM50464P-15. IC part numbers are generally written within the rectangular IC symbol.

Meters Figure 4-21 shows the standard symbol for meters. The type of meter required is identified by a letter code placed within the circular

FIGURE 4-21 Meter symbol. Use letter to identify type of meter.

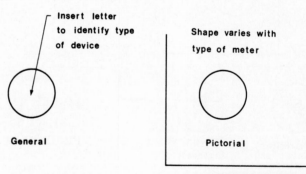

LETTER CODE	Type of Meter
OHM	Ohm
PH	Phase
t°	Temperature
V	Volt
W	Watt
CRO	Oscilloscope
DB	Decibel
F	Frequency
I	Indicating

meter symbol. Figure 4-21 includes a list of some of the more common meter identification letters.

Meters are rated by their operating ranges. For example, voltmeters may have a range of 0 to 10 or 10 to 100 volts. Meters are generally identified on a schematic diagram by range and by a manufacturer's part number.

Resistors   Figure 4-22 shows the standard symbols for resistors. Resistors are used to control current and are rated using ohms. See Appendix B for an explanation of resistor color coding.

Switches   Figure 4-23 show the standard symbols for switches. Switches are rated by the number of poles and available operating positions. For example a double-pole, double-throw (DPDT) switch. Switches are also identified by their operating mechanism, such as rotary or toggle.

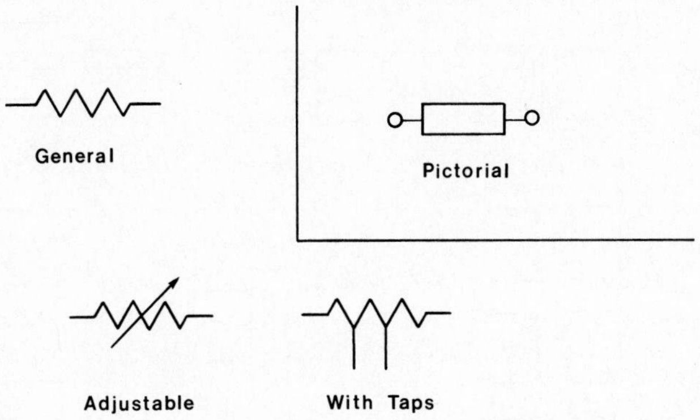

FIGURE 4-22 Resistor symbols.

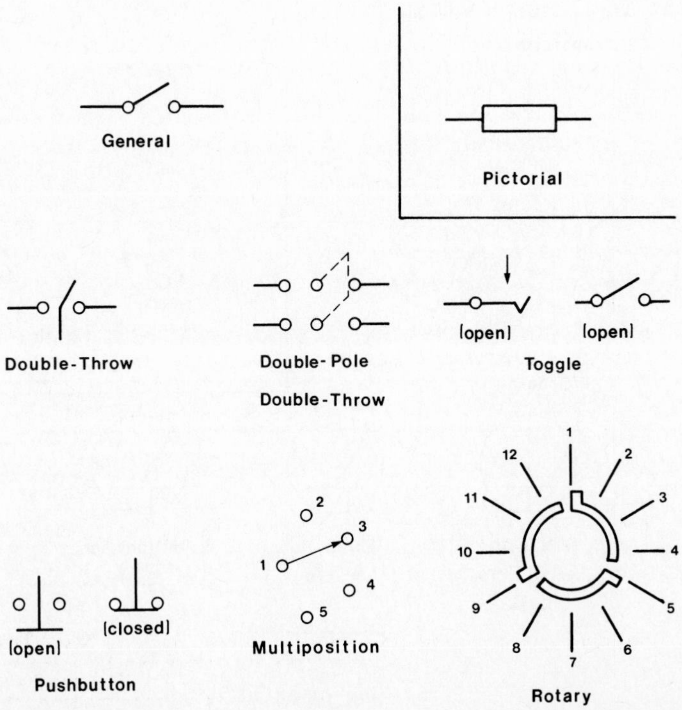

FIGURE 4-23 Switch symbols.

Transformers Figure 4-24 shows the standard symbols for transformers. Transformers are used to change a current value and are rated as to primary current (current coming into the transformer) and secondary current (current coming out of the transformer). They are generally located on the left side of a schematic diagram.

Transistors Figure 4-25 shows the standard symbols for transistors. There are many different types of transistors. They are identified by manufacturer's part number, such as 2N3392. This part number is often included on the schematic next to the transistor.

FIGURE 4-24 Transformer symbols.

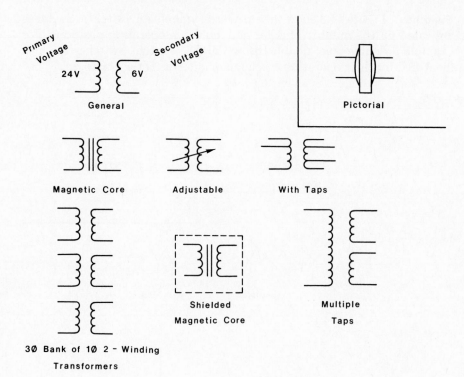

FIGURE 4-25 Transistor symbols.

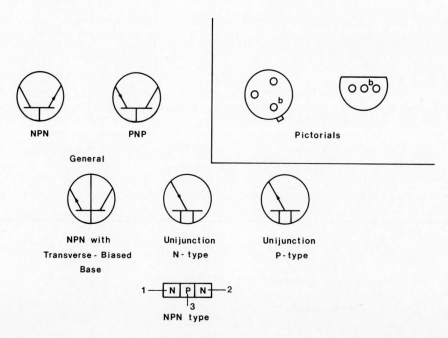

Other symbols   Figure 4-26 shows several other standard symbols commonly used on electronic schematic drawings.

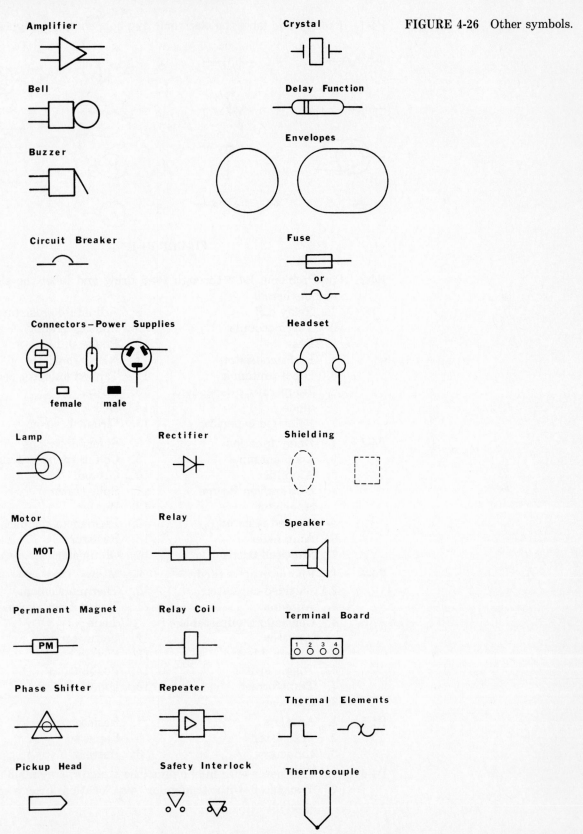

FIGURE 4-26   Other symbols.

Problems

P4-1   Redraw and label the electronic symbols shown in Figure P4-1.

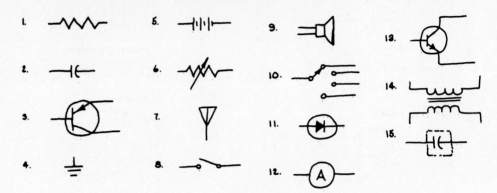

FIGURE P4-1

P4-2   For Problems P4-2 through P4-6 draw and label the electronic symbols listed.

1. Relay coil
2. Thermocouple
3. Fuse
4. PNP transistor
5. Dipole antenna
6. Temperature-dependent diode
7. Polarized capacitor
8. Adjustable resistor
9. Chassis ground
10. Phase shifter
11. NPN transistor
12. Tapped inductor coil
13. Battery
14. Bell
15. Circuit breaker

P4-3
1. Delay function
2. Loop antenna
3. Amplifier
4. Unijunction P-type transistor
5. Tapped resistor
6. Ampmeter
7. Multicell battery
8. Safety interlock
9. Counterpoise antenna
10. Ground
11. Split stator
12. Buzzer
13. Permanent magnet
14. Resistor
15. Ohmmeter

P4-4
1. Photosensitive diode
2. Shielded capacitor
3. Antenna
4. Continuous adjustable inductor
5. Terminal board
6. Toggle switch
7. Transformer
8. Speaker
9. Motor
10. Thermal element
11. Temperature-dependent diode
12. Voltmeter
13. Pickup head
14. Pushbutton
15. Dipole antenna

P4-5
1. Antenna
2. Transistor
3. Inductor
4. Resistor
5. Capacitor
6. Battery

P4-6
1. Transistor with in-line pins (use a pictorial symbol)
2. Transistor with triangular pin location (use a pictorial symbol)

3. General Semiconductor device 0.50 inch × 1.00 inch with five pins located on each of the 1.00-inch sides
4. Amplifier device with five pins
5. Transistor, N-channel type (use a pictorial symbol)
6. Delay function
7. General Semiconductor device 20 mm × 80 mm with four pins on each of the 80-mm sides
8. Diode
9. Chassis ground
10. Ampmeter
11. Amplifier devices with three pins
12. Phase shifter
13. A General Semiconductor device 1.50 inches × 3.50 inches. Locate 16 equally spaced pins on each of the 3.50-inch sides. Number of pins from 1 through 16 on the left side from top to bottom and 17 through 32 on the right side from top to bottom.

5

Schematic Diagrams

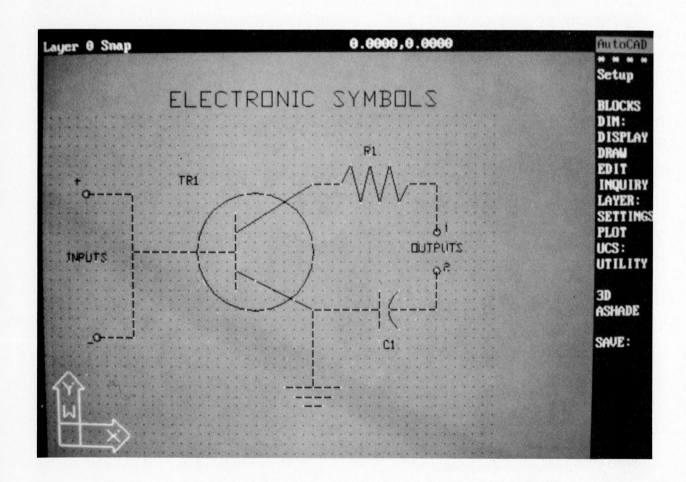

5-1   INTRODUCTION

Schematic diagrams are drawings that show graphically what components are to be used and how these components are to be connected to form a desired circuit. There are two basic types of schematic diagrams: those which use electronic symbols and those which use pictorial symbols. In this chapter we present both types of diagrams.

5-2   HOW TO DRAW SCHEMATIC DIAGRAMS

When drawing schematic diagrams, it is important that the drawing be easy to follow. The symbols should be presented clearly and in an organized format. All relative data, such as component values and identification numbers, should be located as close as possible to the appropriate symbols [Figure 5-1(a)].

To lay out the diagram clearly, the individual symbols should not be cramped together. A crowded group of symbols is difficult to read and is therefore more likely to cause error. Cramping can be avoided by planning the placement of the symbols ahead of time.

One method that may be used to help assure neat, uncluttered schematic diagrams is to use a grid background (see Section 4-3). The grid can be used to position and align components and conductor paths. Figure 5-2 shows five ICs positioned and aligned using a point background. The ZOOM function was used to enlarge the areas between the components so that the area for conductor paths can be more easily seen.

Figure 5-3 shows the same components presented in Figure 5-2 with conductor paths, pin labels, and components numbers added. The conductor paths were added using the POINT-TO-POINT function. The points on the grid background are known to the computer, so can be easily used to create lines.

Component values and/or stock numbers may be included by lettering the data next to the component as shown in Figure 5-4 or by setting up a parts list as shown in Figure 5-1(b). Parts lists have the advantage of allowing purchasers to work from a specific list of components rather than creating their own lists from the schematic. Including the data next to the symbols allows technicians to read the diagram more quickly, without having to refer constantly to the table.

When a parts list is used with a schematic diagram, coded part numbers are assigned to each component. For example, all resistors are assigned the letter R followed by a number. The numbers start with 1 and continue as necessary. Figure 5-5 shows a listing of some of the most

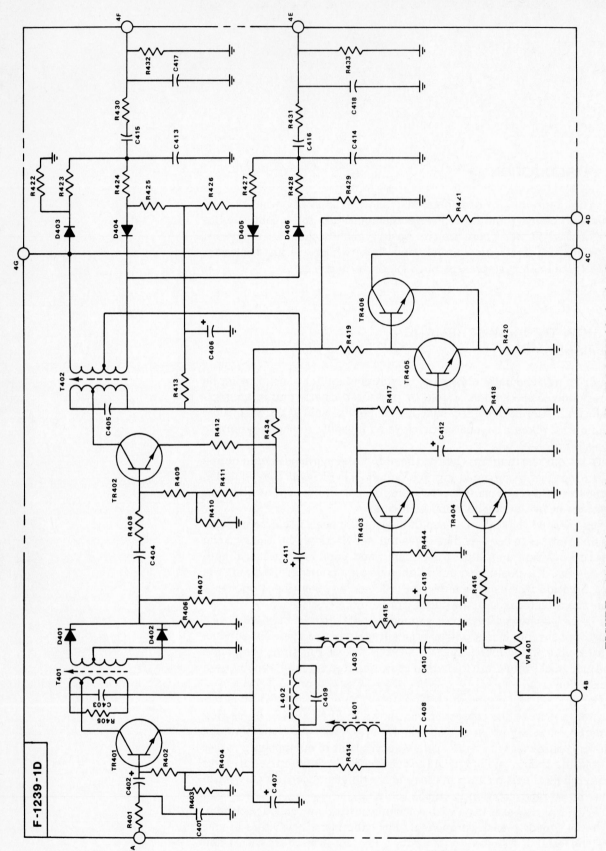

FIGURE 5-1(a) Schematic diagram for the FM MPX block of a Sansui 350A solid-state AM/FM stereo tuner amplifier.

PARTS LIST– F-1239-1D

Part No	Value / Name			Part No	Value / Name			
R401	1 k			C412	1 UF	50V	EL	
R402	100 k			C413	680 PF	±5% 50V	ST	
R403	15 k			C414	680 PF			
R404	22 k			C415	0.15 UF	±10% 50V	MY	
R405	68 k			C416	0.15 UF			
R406	100 k			C417	0.006 UF	±5% 50V	MY	
R407	100 k			C418	0.006 UF			
R408	4.7 k			C419	1 UF	50	EL	
R409	100 k							
R410	2.2 k							
R411	22 k							
R412	33 k			TR401	2SC711			
R413	220 k			TR402	2SC711			
R414	47 k			TR403	2SC711			
R415	2.2 k	ALL VALUES ARE		TR404	2SC711			
R416	47 k	IN OHMS		TR405	2SC733			
R417	22 k	±10% $\frac{1}{4}$W CB		TR406	2SC735			
R418	22 k							
R419	3.3 k							
R420	4.7							
R421	47			D401	IN34A			
R422	220 k			D402	IN34A			
R423	10 k			D403	IN34A			
R424	10 k			D404	IN34A			
R425	220 k			D405	IN34A			
R426	220 k			D406	IN34A			
R427	10 k							
R428	10 k							
R429	220 k							
R430	56 k			T401	19 kHz Coil			
R431	56 k			T402	38 kHz Coil			
R432	15 k							
R433	15 k							
R434	47 k							
				L401	19 kHz Coil			
				L402	Micro Inductor			
				L403	67 kHz Coil			
C401	68 PF	±10% 50V	CE					
C402	10 UF	10V	EL					
C403	10000 PF	± 5% 50V	ST					
C404	0.022 UF	±10% 50V	MY					
C405	4700 PF	± 5% 50V	ST					
C406	1 UF	50 V	EL					
C407	47 UF	25 V	EL					
C408	10000 PF							
C409	2200 PF	± 5% 50V	ST					
C410	270 PF							
C411	10 UF	25 V	EL					

FIGURE 5-1(b) Schematic diagram parts list for the Sansui 350A shown in part (a). (Courtesy of Sansui Electric Co., Japan.)

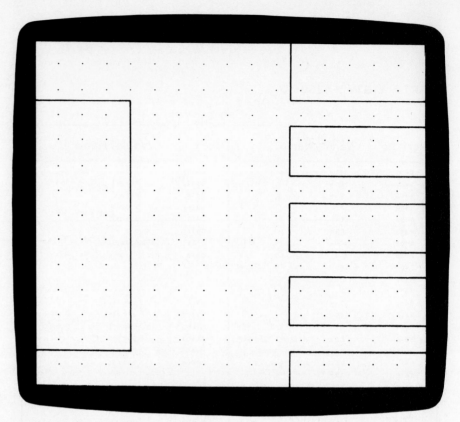

FIGURE 5-2 Symbols aligned using a dot grid pattern.

FIGURE 5-3 The symbols presented in Figure 5-2 with pin numbers and conductor paths added.

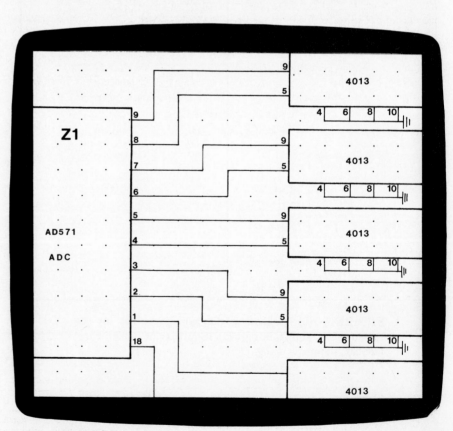

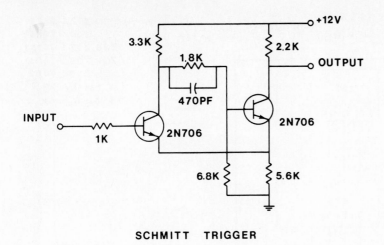

SCHMITT TRIGGER

FIGURE 5-4   Schematic diagram.

COMPONENT	CODE
Capacitor	C
Diode	D
Induction Coil	L
Resistor	R
Switch	S
Transformer	T
Transistor	TR or Q
Integrated Circuit	Z

FIGURE 5-5   Component letter codes.

widely used component letter codes. Figure 5-6 shows an example of a schematic diagram together with an appropriate parts list.

Note that both diodes are labeled D1. This means that both diodes are exactly the same component. All equivalent components are labeled identically. If, for example, a circuit contained seven 15-kilohm resistors, all would be labeled R with the same number.

Many smaller schematic diagrams do not require parts lists. The value data and manufacturers' numbers are either written directly on the schematic next to the appropriate component or, as shown in Figure 5-7, are coded by including the code interpretations on the body of the drawing. In Figure 5-7, the code interpretations are included in the upper right-hand corner of the schematic diagram.

Schematic diagrams should be drawn in a neat, organized manner. If possible, connector lines should be evenly spaced and major components should be located so that connector lines can be drawn directly (straight) between terminals. This will not only give the drawing a pleasant appearance, but will make it easy to read.

FIGURE 5-6   Schematic diagram with separate parts list.

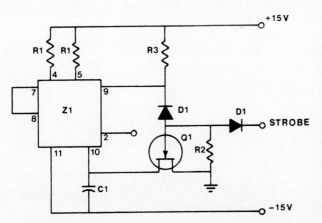

NO.	VALUE/ NAME	QTY
Z1	ICL8083	1
Q1	2N4392	1
D1	1N914	2
C1	3300 pF	1
R3	15KΩ ± 10% $\frac{1}{4}$ W	1
R2	100KΩ ± 10% $\frac{1}{4}$ W	1
R1	10KΩ ±10% $\frac{1}{4}$ W	2
NO.	VALUE/ NAME	QTY

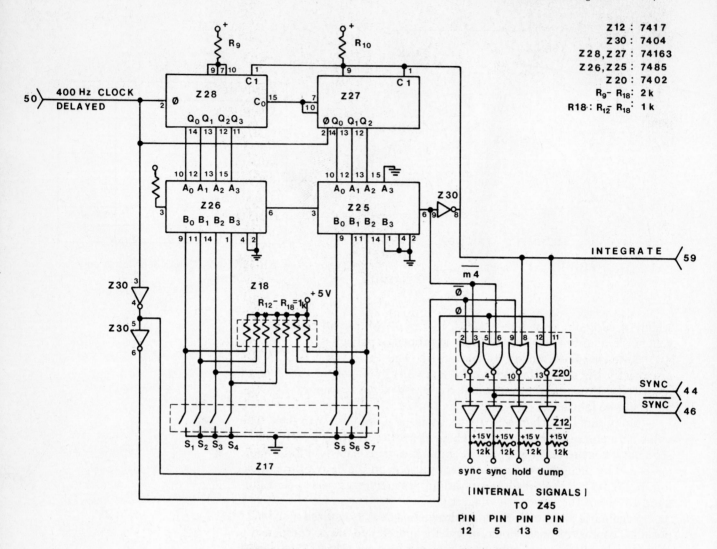

FIGURE 5-7   Schematic diagram parts list included.

Figure 5-8 shows an example of a poorly laid out schematic diagram. The spacing is uneven and the connector lines ramble. Figure 5-8 also shows a corrected version in which the lines are evenly spaced and direct.

Schematic diagrams should be drawn following the following general guidelines.

Left side — inputs

Right side — outputs

Lower area — grounds

Lower left — power

In general, schematic diagrams are read from left to right and from top to bottom, as shown in Figure 5-9.

If possible, arrange the schematic diagram so that it can be read from left to right in terms of function. Place together on the drawing symbols that work together to perform a function.

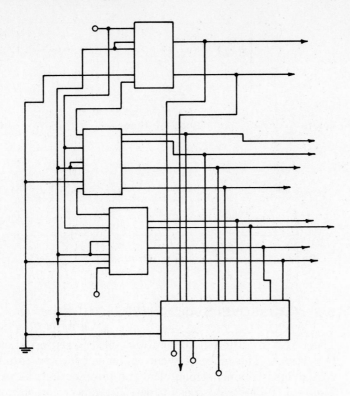

A poorly laid out schematic

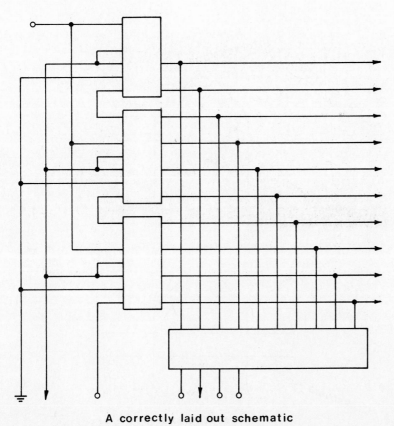

A correctly laid out schematic

FIGURE 5-8   How to lay out a schematic diagram.

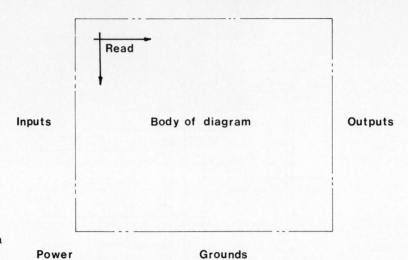

FIGURE 5-9 How to read a schematic diagram.

5-3 CROSSOVERS AND INTERSECTIONS

There are two different conventions used to represent crossovers and intersections. The older convention, which is now rarely used, pictures crossovers as loops (semicircles) and intersections as two lines crossing. Figure 5-10 illustrates both types. The newer convention is the dot convention, which indicates intersections with dots and crossovers by crossed lines (see Figure 5-10). With the dot system, when a line joins a line at 90° to form an obvious intersection, the dot may be omitted.

FIGURE 5-10 How to draw crossovers and intersections on a schematic diagram.

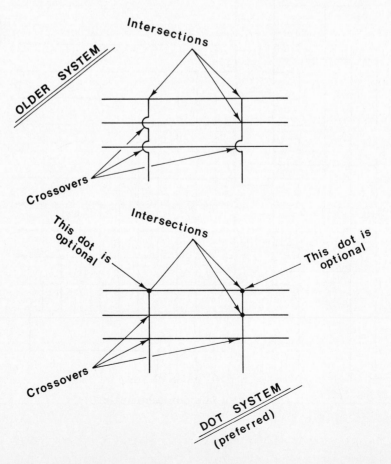

5-4   BASIC CIRCUITS

Figure 5-11 illustrates some of the more common circuit configurations. These basic configurations are found as part of many more sophisticated circuit designs. The drafter should be familiar enough with these basic circuits to recognize them when they appear in design sketches.

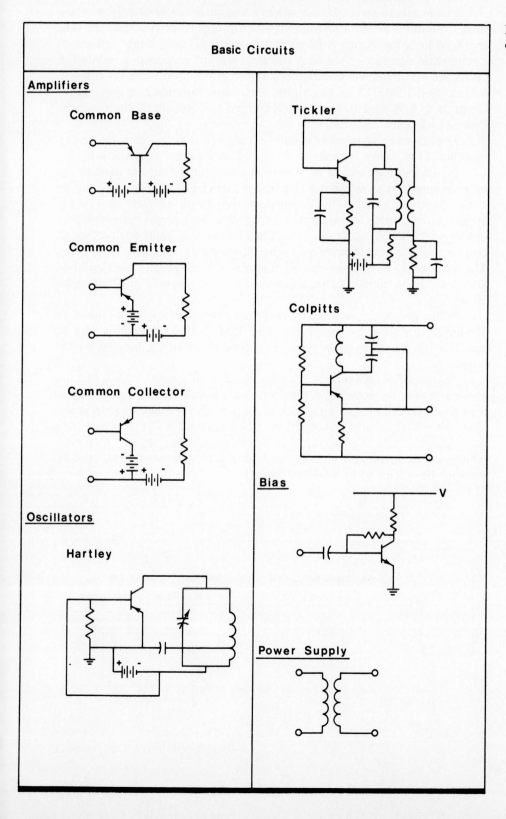

FIGURE 5-11   Some basic circuits.

5-5 INTEGRATED CIRCUITS
IN SCHEMATIC DIAGRAMS

Integrated circuits (ICs) are used extensively in many different types of schematic diagrams. They usually appear as rectangles, squares, or equilateral triangles. See Section 4-4 for an explanation of the graphic symbols used to represent ICs.

Pin numbers on an IC are always assigned consecutively starting with 1. Figure 5-12 shows a dual-type D flip-flop IC device with the pins arranged in order. Figure 5-12 shows the same IC as it might appear on a schematic diagram. Note that the pins are not arranged in numerical order but are arranged by function. Input pins 5 and 9 are on the left, output pins 1 and 13 on the right, and other functions in the middle. Grounds 4, 6, 8, and 10 are located toward the bottom of the diagram. Pins not used are omitted.

Designers and engineers usually arrange IC pins to follow the design function. This makes it easier to read and understand the diagram.

IC devices may also be arranged on a schematic diagram to show the internal function as related to the rest of the circuit. Figure 5-13 shows a wide-bandwidth dual-JFET-input operation amplifier: first as it would appear with the pins in consecutive order, then as it might appear as part of a circuit on a schematic diagram. The pins are still numbered, although two amplifier symbols are used rather than the rectangular IC symbol. The notation ½ 353 states the IC device's manufacturer's part number, 353, with the ½ implying that the amplifier shown is only one-half of the IC device.

When preparing schematic diagrams, always arrange the pins of an IC device to follow the design function. The shorter edges may also be used for pin locations even though no pins are actually located on the edges.

IC amplifier devices are manufactured in different shapes, but are always drawn on schematic diagrams as equilateral triangles. Figure 5-14 shows a JFET-input operation amplifier. The actual amp chip is rectangular, with four pins on each of the two longer sides. A manufacturer's description of the chip would also be rectangular as shown. However, when used as a triangle, as shown in Figure 5-14, pins are represented by straight lines located around the triangle.

FIGURE 5-12 Not all pins need be used or drawn in consecutive order.

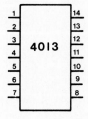

As manufactured

As could appear on

a schematic diagram

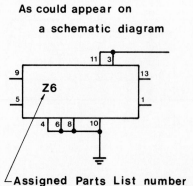

Assigned Parts List number

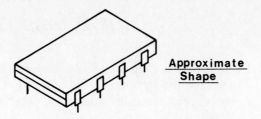

Approximate
Shape

Pin Connection

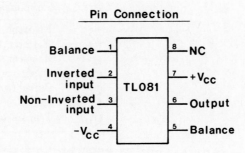

As represented on a schematic diagram

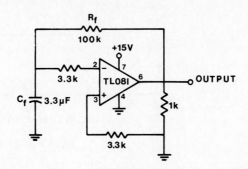

FIGURE 5-13 Sample schematic diagram that uses an IC.

FIGURE 5-14   Split IC as drawn in a schematic diagram.

As manufactured

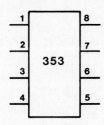

As could appear on a schematic diagram

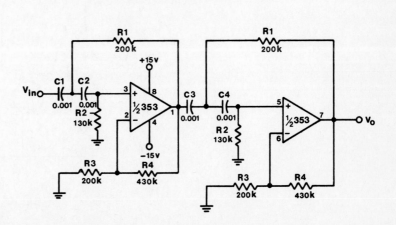

Problems

P5-1   Redraw the Colpitts oscillator circuit shown in Figure P5-1 and replace the letters R1, R2, R3, C1, C2, L1, and Q1 with the appropriate symbol and value. The letters are defined as follows:

Letter	Component	Value
R1	Resistor	12 KΩ
R2	Resistor	8.2 KΩ
R3	Resistor	1.5 KΩ
C1	Capacitor	0.10 PF
C2	Capacitor	0.047 PF
L1	Inductor	10 MH
Q1	Transistor	2N2926

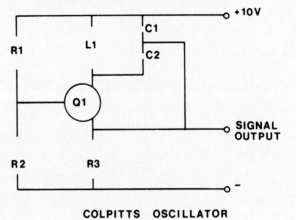

COLPITTS OSCILLATOR

FIGURE P5-1

P5-2   Redraw the basic logic circuit shown in Figure P5-2 and add the following values.

R1   10KΩ
R2   4.7 KΩ
Q1   2N708   (NPN transistor)

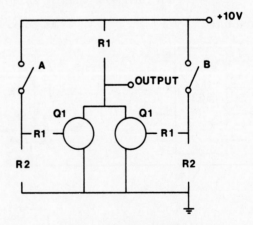

BASIC LOGIC CIRCUIT

FIGURE P5-2

P5-3   Redraw the power supply circuit shown in Figure P5-3.

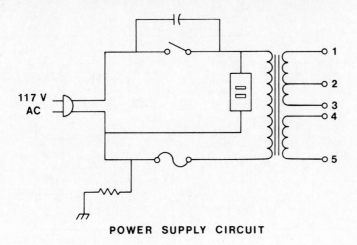

POWER SUPPLY CIRCUIT

FIGURE P5-3

P5-4   Redraw the AM broadcast band transmitter shown in Figure P5-4 and substitute as follows:

1. 670 KC CRYSTAL	11. 1 KΩ
2. 150 KΩ	12. Battery 6 V
3. 2N170 (NPN transistor)	13. Ground
	14. 0.001 PF
4. 300 PF	15. 0.003 PF
5. Ground	16. L3
6. Ground	17. Ground
7. 2N170 (NPN transistor)	18. 10 mF, 10 V
	19. 0.001 PF
8. 0.002 PF	20. Antenna
9. Ground	21. Chassis ground
10. Inductor (10 MH RFC)	

Add both symbols and values.

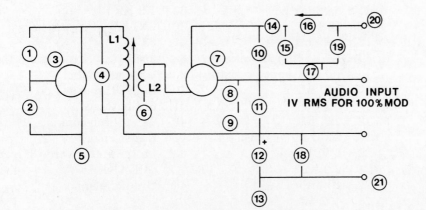

AM BROADCAST BAND TRANSMITTER

FIGURE P5-4

P5-5   Redraw the schematic diagram for the 6-volt phono amplifier
shown in Figure P5-5 and substitute as follows:

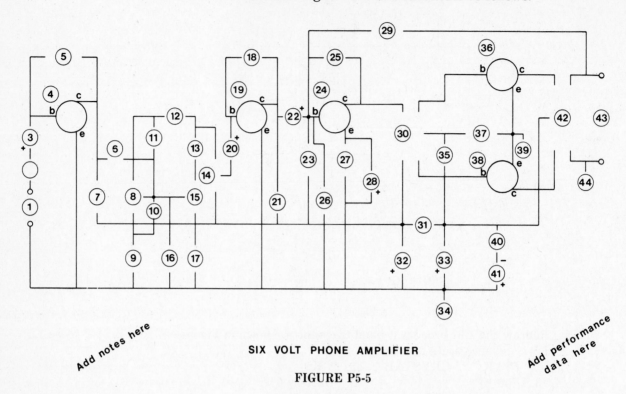

SIX VOLT PHONE AMPLIFIER

FIGURE P5-5

 1. Letter in the words   24. 2N323   (transistor)
    CRYSTAL   25. 47 KΩ
    CARTRIDGE   26. 1.5 KΩ
 2. 10 KΩ   27. 330 Ω
 3. 8 mF   28. 6 V, 50 mF
 4. 2N323   (transistor)   29. 33 KΩ
 5. 150 KΩ   30. T1   (iron-core
 6. 0.5 mF   transformer)
 7. 6.8 KΩ   31. 220 Ω
 8. R1   (tapped resistor)   32. 6 V, 50 mF
 9. 1 KΩ   33. 6 V, 50 mF
10. 0.2 mF   34. Ground symbol
11. 0.02 mF   35. 1.2 KΩ
12. 10 KΩ   36. 2N1415   (transistor)
13. 0.05 mF   37. 33 Ω
14. R3   (tapped resistor)   38. 2N1415   (transistor)
15. R2   (tapped resistor)   39. Ground symbol
16. 100 KΩ   40. Single-throw switch
17. 0.1 mF   41. 6-V battery
18. 220 KΩ   42. T2   (iron-core
19. 2N323   (transistor)   transformer)
20. 6 V, 10 mF   43. Letter in the words TO
21. 2.2 KΩ   SPEAKER
22. 10 mF   44. Ground symbol
23. 4.7 KΩ

All transistors are PNP types. In addition, add the following
notes and performance data.

Notes

R1 — Bass control — 50 K Linear taper
R2 — Treble control — 50 K Linear taper
R3 — Volume control — 10 K Audio taper
T1 — Driver Transformer — PR1 2 K/sec 1.5 K, C.T.
T2 — Output transformer — PR1 100Ω/sec V.C.(3.2, 8, 16Ω)
All resistors ½ watt

Performance Data

Maximum power out at 10% distortion	300 mW
Distortion at 100 mW	60 Hz — 3%
	1.0 RHz — 1.5%
	5.0 RHz — 3.0%

P5-6 Much of the work assigned beginning drafters is in the form of freehand sketches. An engineer will present freehand design sketches and have a drafter prepare the finished schematic drawings from these sketches. Study the design sketches in Figure P5-6 and redraw them using drawing instruments.

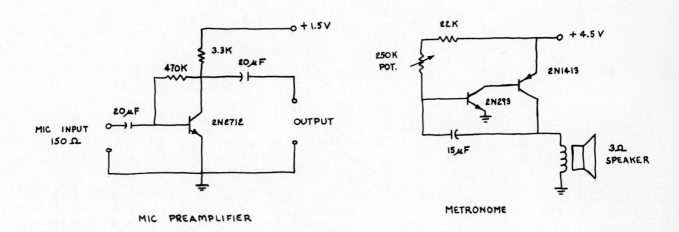

MIC PREAMPLIFIER   METRONOME

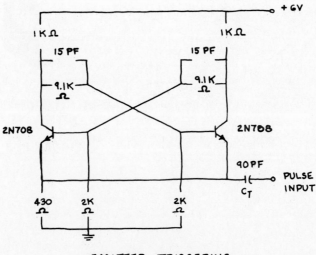

EMITTER TRIGGERING

FIGURE P5-6

P5-7 Redraw the pictorial schematics in Figures P5-7 through
through P5-11 as schematic diagrams using symbols.
P5-11

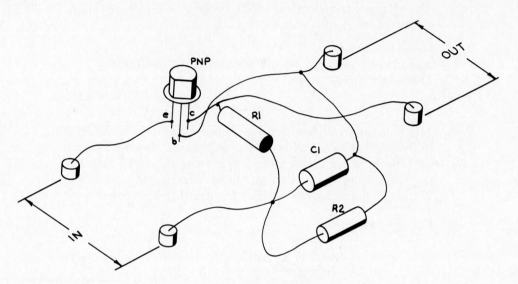

FIGURE P5-7

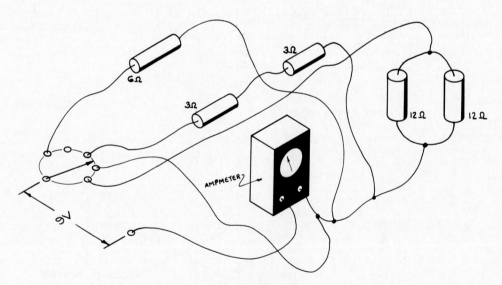

FIGURE P5-8

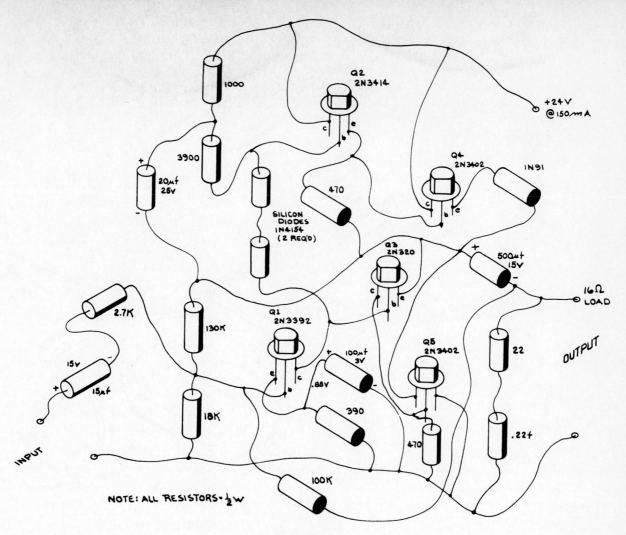

FIGURE P5-9

Labels within figure: 1000, Q2 2N3414, c e b, +24V @150mA, 3900, +20μf 25V -, 470, Q4 2N3402, c b e, 1N91, SILICON DIODES 1N4154 (2 REQ'D), Q3 2N320, c b e, +500μf 15V -, 16Ω LOAD, 2.7K, 130K, Q1 2N3392, +100μf 3V, .85V, Q5 2N3402, 22, OUTPUT, 15V, 15μf +, e c b, 390, 470, .22f, 18K, INPUT, 100K, NOTE: ALL RESISTORS = ½ W

FIGURE P5-10

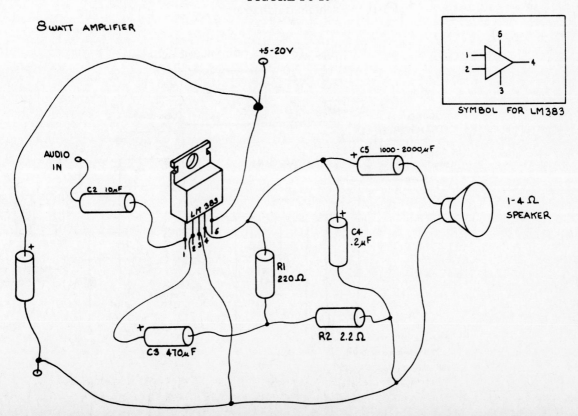

8 WATT AMPLIFIER

Labels within figure: +5-20V, SYMBOL FOR LM383, AUDIO IN, C2 10μF, LM383, C5 1000-2000μF, 1-4Ω SPEAKER, C4 .2μF, R1 220Ω, 1 2 3 4, C3 470μF, R2 2.2Ω

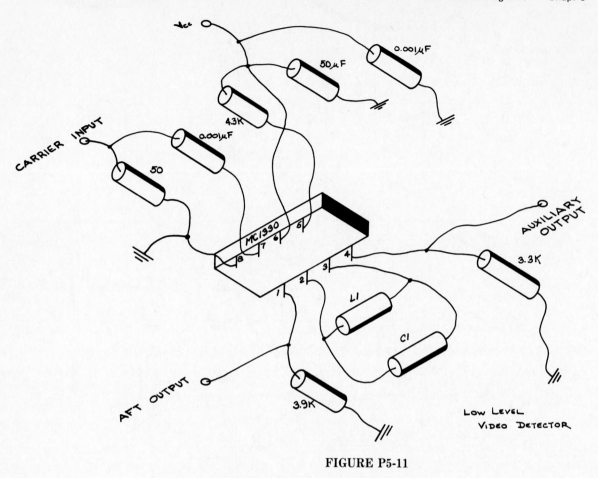

FIGURE P5-11

P5-12   Figures P5-12 through P5-16 are designers' sketches of var-
through ious circuits. In each problem, the sketch is to be redrawn as
P5-16   a schematic diagram.

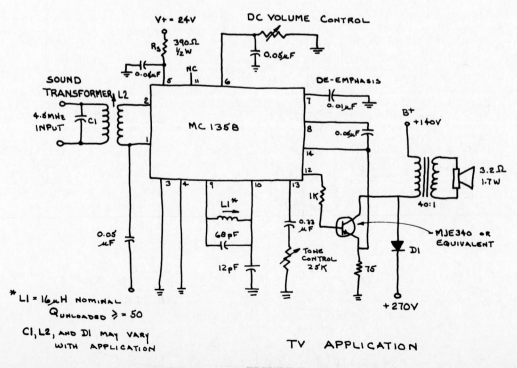

FIGURE P5-12

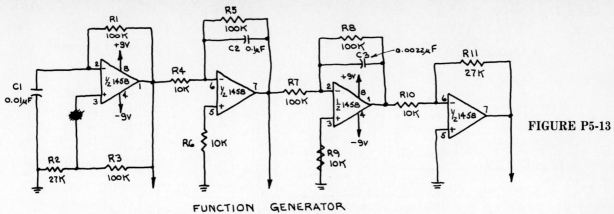

FUNCTION GENERATOR

FIGURE P5-13

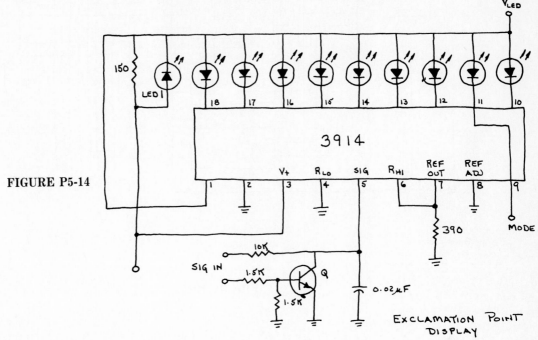

FIGURE P5-14

EXCLAMATION POINT
DISPLAY

FIGURE P5-15

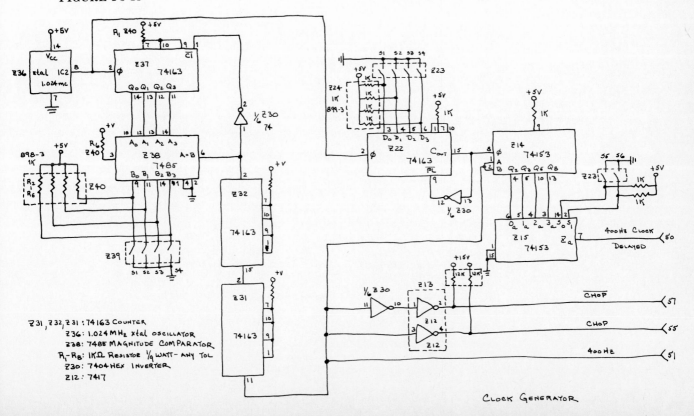

Z31, Z32, Z31 : 74163 COUNTER
Z36: 1.024 MHz XTAL OSCILLATOR
Z38: 7485 MAGNITUDE COMPARATOR
R_1-R_8: 1KΩ RESISTOR 1/4 WATT - ANY TOL
Z30: 7404 HEX INVERTER
Z12: 7417

CLOCK GENERATOR

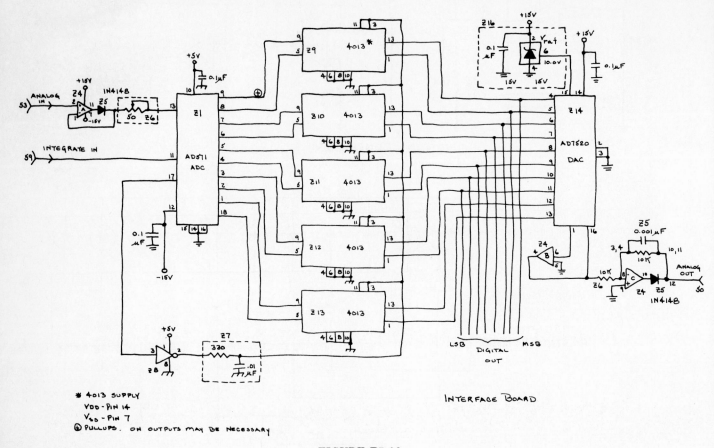

FIGURE P5-16

6

Printed Circuit Diagrams

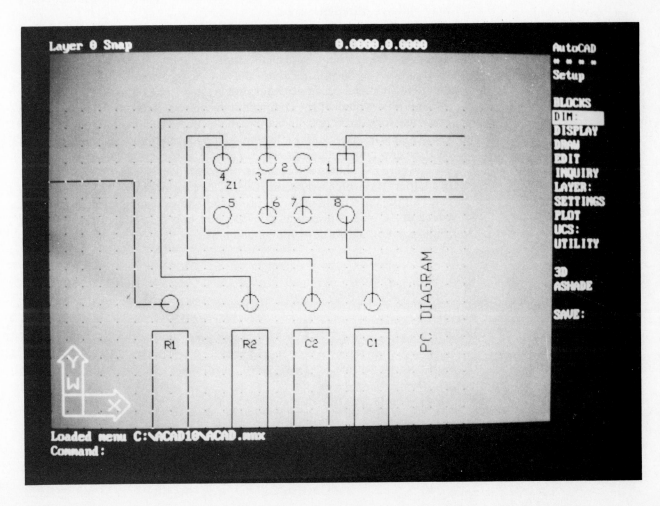

6-1 INTRODUCTION

In this chapter we study how to prepare the various drawings involved in the design and manufacturing of printed circuits (PCs). These drawings include freehand design sketches, space allocation layouts, conductor paths, taping, solder masks, and drill drawings.

6-2 WORKING FROM SCHEMATICS

PC drawings are created from schematic diagrams such as the one shown in Figure 6-1. As we learned from Chapter 5, schematic diagrams are design drawings that show the function of a circuit. They do not necessarily have any relationship to the physical size and location of the actual components. For example, all resistors use the same symbol even though they may all have different sizes.

The first step in creating a set of PC drawings is to change the initial schematic diagram from one that uses representative symbols to one that uses pictorial symbols. This can initially be done using a freehand sketch such as the one shown in Figure 6-2.

The pictorial schematic diagram is used to help determine the number and type (value) of the components required. Some drafters/designers prefer to work directly from the original schematic, but sketching a pictorial schematic takes very little time and helps to maintain the functional relationship between components.

Figure 6-3 shows a schematic diagram [part (a)] and a pictorial diagram [part (b)] which utilize an IC chip. Not all the pins are needed; however, all pins are numbered. Figure 6-4(a) shows a schematic diagram in which an IC chip is drawn as two different halves. The pictorial diagram [Figure 6-4(b)] shows the chip as one unit.

FIGURE 6-1 Schematic diagram.

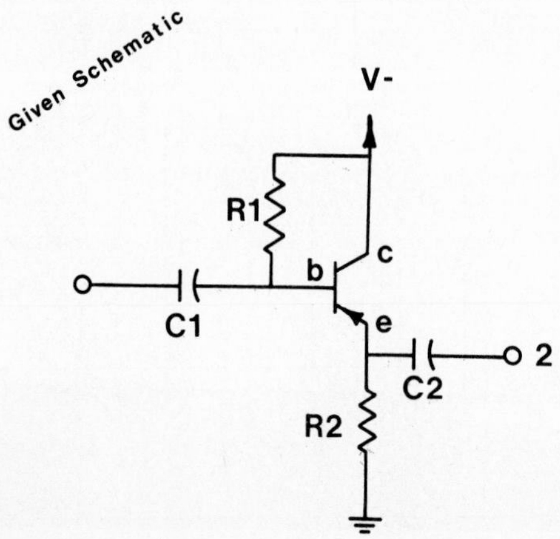

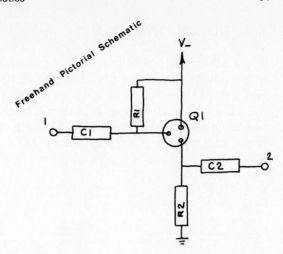

FIGURE 6-2 Freehand pictorial sketch of the schematic diagram in Figure 6-1.

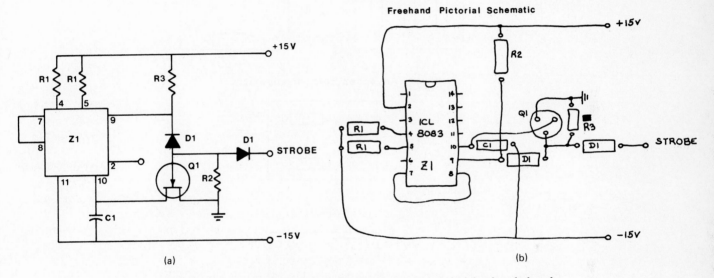

FIGURE 6-3 (a) Schematic diagram and (b) its pictorial freehand sketch.

FIGURE 6-4 (a) Schematic diagram and (b) its pictorial freehand sketch.

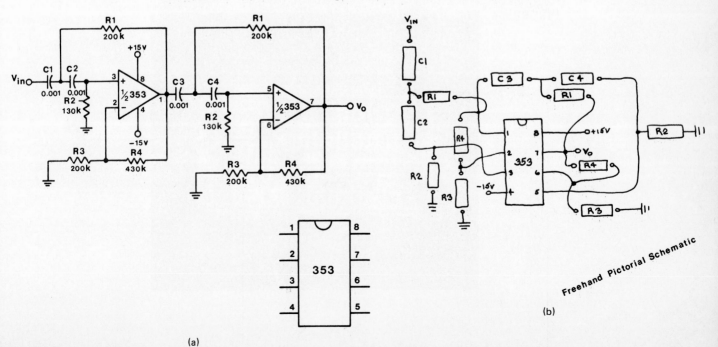

6-3 DESIGN PRINCIPLES

After converting a schematic into a pictorial schematic, the components and conductor paths must be arranged on the board. Several design principles must be considered when arranging the components.

If possible, all components should be aligned in a common direction, as shown in Figures 6-5 and 6-6. Uniform alignment helps simplify PC board manufacture. Keep conductor paths as short as possible, and avoid sharp turns and acute angles, as shown in Figure 6-7.

Best   arrangement

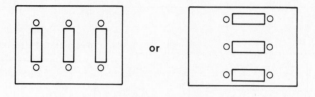

or

An acceptable arrangement

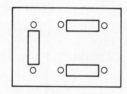

Avoid !!

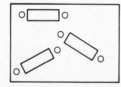

FIGURE 6-5   How to lay out a PC diagram.

FIGURE 6-6

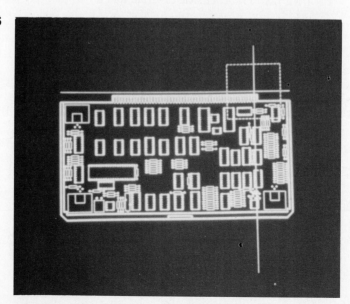

CONDUCTOR PATHS

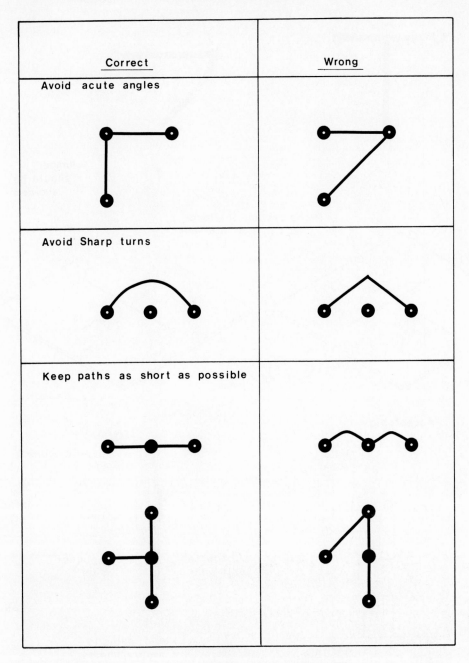

FIGURE 6-7   Keep all conductor paths as short as possible and avoid sharp turns.

If possible, make all conductor path intersections on pads, as shown in Figure 6-8. Intersections made directly into conductor paths tend to develop microscopic cracks and thereby cause poor reliability.

Conductor paths should be arranged to minimize crossovers, as shown in Figure 6-9. Conductor paths on PC boards are not shielded so cannot crossover each other as can insulated wires. When conductor paths intersect, one of the paths must pass through the board to the opposite side, pass under the other conductor path, then return to its original side and resume its path. This procedure, called a "jumper," adds time and expense to the PC board's manufacture and should be avoided.

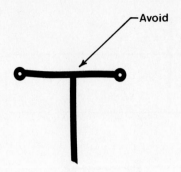

—Avoid

—Make intersections
on pads

FIGURE 6-8 Avoid T connections.

FIGURE 6-9 Avoid crossovers.

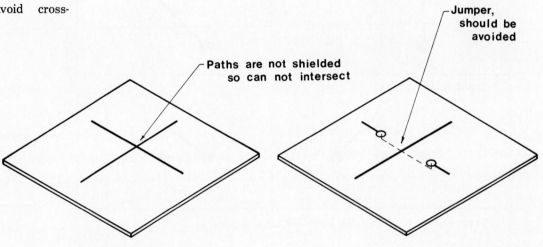

Paths are not shielded
so can not intersect

Jumper,
should be
avoided

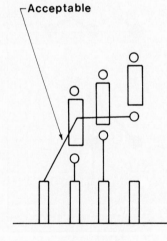

Acceptable

Unacceptable

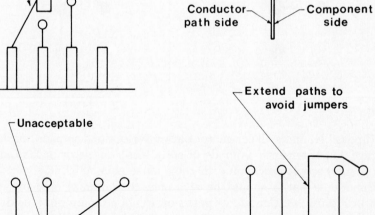

Component

Path

Conductor
path side

Component
side

Extend paths to
avoid jumpers

Figure 6-10 shows how these principles are applied to the schematic diagram presented in Figure 6-2. All inputs and outputs were located on the same board edge; then several attempts were made to align the components, minimize conductor paths, avoid jumpers, and so on. Note how different the third and final arrangement is from the original pictorial schematic. Figure 6-11 shows how the same procedure might be done using a computer. Figures 6-12 and 6-13 show similar sketches for the schematics shown in Figures 6-3 and 6-4.

Figure 6-14 shows a design layout done using a computer. The components were located and aligned, then the conductor paths added. The final arrangement of the conductor paths is accomplished by reworking the layout directly on the screen. This procedure is called *editing*.

Figure 6-15 shows another design layout done using a computer. Note that the conductor paths are either horizontal, vertical, or 45° slanted lines. There are no curved lines. Straight or slanted line conductor paths are acceptable alternatives to curved paths.

FIGURE 6-10   Arranging a PC board layout.

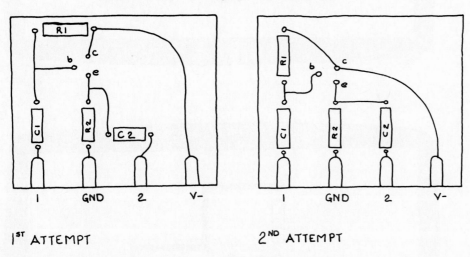

1ˢᵀ ATTEMPT   2ᴺᴰ ATTEMPT

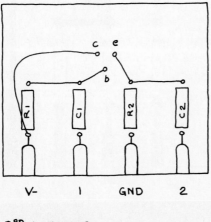

3ᴿᴰ ATTEMPT

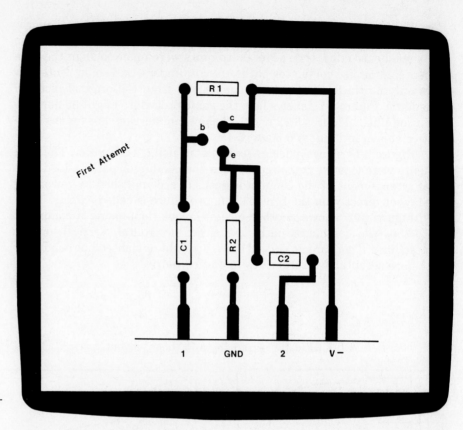

FIGURE 6-11 A PC board design done using a computer.

FIGURE 6-11 (*cont'd*)

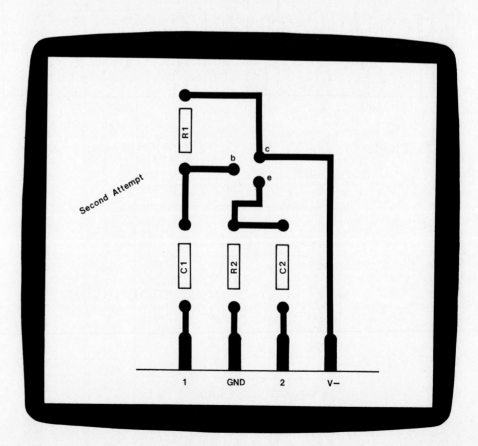

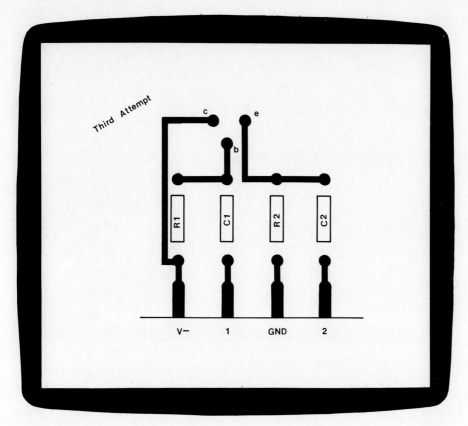

FIGURE 6-11 (*cont'd*)

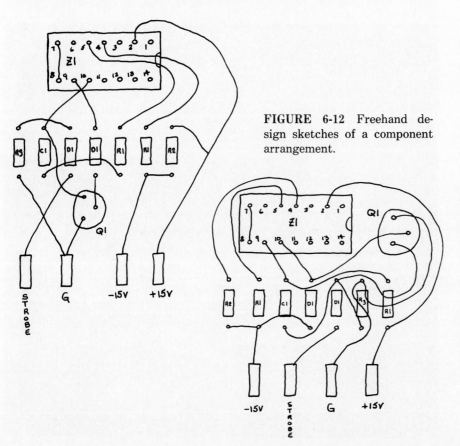

FIGURE 6-12 Freehand design sketches of a component arrangement.

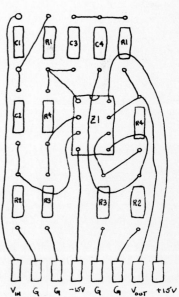

FIGURE 6-13 Freehand design sketches of a component arrangement.

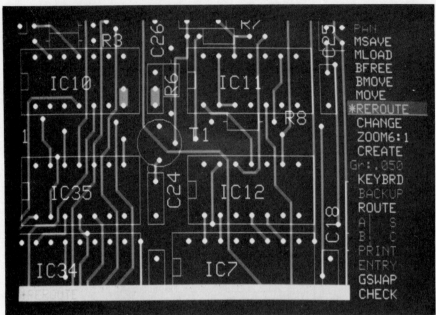

FIGURE 6-14   A sample design layout. (Courtesy of Visionics Corp., Sunnyvale, Calif.)

FIGURE 6-15   A design layout.

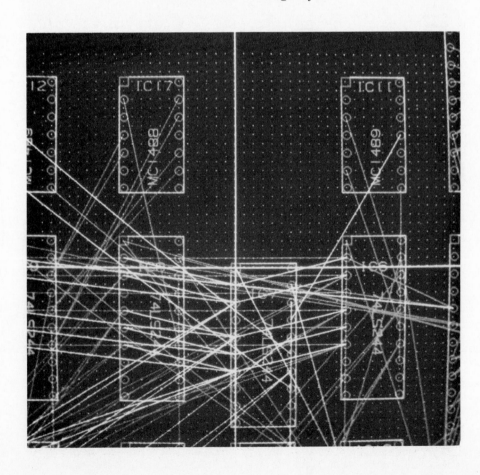

6-4   SPACE ALLOCATION LAYOUTS (MASTER LAYOUTS)

After an acceptable component configuration has been sketched, the size of components is considered. It should be pointed out that after the component sizes are considered, the sketches must sometimes be reworked. Component sizes may vary considerably, so a back-and-forth process between arrangement sketches and space allocation layout is not unusual.

The actual size of components can be found in manufacturers' catalogs. Figure 6-16 shows a sample page from a Mallory catalog. Note that the page includes information as to the components' performance and capacity as well as size data. Appendix H lists some other component sizes. Component sizes may be listed in either English or metric units.

The mounting requirement of each component is also an important consideration. Space must be alloted not just for the component, but also for the area needed to mount the component to the board. For example,

FIGURE 6-16   Sample manufacturer's catalog. (Courtesy of Mallory.)

Film Capacitors/Type M192P

Polyester, Axial Leaded

M192P capacitors are wound with a polyester film dielectric and thin gage foil under carefully controlled conditions, permitting reliable working voltages of 80, 200, and 400 volts. The capacitor features extended foil sections which are terminated in metal end caps, assuring a consistent capacitor size as well as a non-inductive connection. The end caps further protect against entry of moisture into the capacitor. Physical sizes of the M192P capacitor have been chosen as closely as possible to those of various composition resistors for compatibility with automatic insertion equipment. **Request bulletin 9-781 for complete specifications. For pricing, reference price sheet No. 430.**

Highlights
Capacitance: .0001μF to .39μF
Voltage: 80, 200 and 400 WVDC
Tolerance: ±10%
Temperature: -55°C to +85°C

Capacitance μF	Diameter	Length	Catalog Number
80 WVDC			
.0022	0.138	0.312	M192P2229R8
.0027	0.138	0.312	M192P2729R8
.0033	0.138	0.312	M192P3329R8
.0039	0.138	0.312	M192P3929R8
.0047	0.138	0.312	M192P4729R8
.0056	0.170	0.375	M192P5629R8
.0068	0.170	0.375	M192P6829R8
.0082	0.170	0.375	M192P8229R8
.01	0.170	0.375	M192P1039R8
.012	0.170	0.437	M192P1239R8
.015	0.170	0.437	M192P1539R8
.018	0.170	0.500	M192P1839R8
.022	0.170	0.500	M192P2239R8
.027	0.204	0.437	M192P2739R8
.033	0.204	0.437	M192P3339R8
.039	0.204	0.562	M192P3939R8
.047	0.204	0.562	M192P4739R8
.056	0.225	0.562	M192P5639R8
.068	0.225	0.562	M192P6839R8
.082	0.290	0.500	M192P8239R8
.1	0.290	0.500	M192P1049R8
.12	0.290	0.625	M192P1249R8
.15	0.290	0.625	M192P1549R8
.22	0.290	0.625	M192P2249R8
.27	0.318	0.875	M192P2749R8
.33	0.318	0.875	M192P3349R8
.39	0.318	1.187	M192P3949R8

Capacitance μF	Diameter	Length	Catalog Number
200 WVDC			
.0001	0.138	0.312	M192P10192
.00012	0.138	0.312	M192P12192
.00015	0.138	0.312	M192P15192
.00018	0.138	0.312	M192P18192
.00022	0.138	0.312	M192P22192
.00027	0.138	0.312	M192P27192
.00033	0.138	0.312	M192P33192
.00039	0.138	0.312	M192P39192
.00047	0.138	0.312	M192P47192
.00056	0.138	0.312	M192P56192
.00068	0.138	0.312	M192P68192
.00082	0.138	0.312	M192P82192
.001	0.138	0.312	M192P10292
.0012	0.138	0.312	M192P12292
.0015	0.138	0.312	M192P15292
.0018	0.170	0.312	M192P18292
.0022	0.170	0.312	M192P22292
.0027	0.170	0.312	M192P27292
.0033	0.170	0.312	M192P33292
.0039	0.170	0.437	M192P39292
.0047	0.170	0.437	M192P47292
.0056	0.170	0.437	M192P56292
.0068	0.170	0.437	M192P68292
.0082	0.170	0.500	M192P82292
.01	0.170	0.500	M192P10392
.012	0.204	0.437	M192P12392
.015	0.204	0.437	M192P15392

Capacitance μF	Diameter	Length	Catalog Number
400 WVDC			
.00047	0.170	0.375	M192P47194
.00056	0.170	0.375	M192P56194
.00068	0.170	0.375	M192P68194
.00082	0.170	0.375	M192P82194
.001	0.170	0.375	M192P10294
.0012	0.170	0.375	M192P12294
.0015	0.170	0.375	M192P15294
.0018	0.170	0.437	M192P18294
.0022	0.170	0.437	M192P22294
.0027	0.204	0.437	M192P27294
.0033	0.204	0.437	M192P33294
.0039	0.204	0.437	M192P39294
.0047	0.204	0.437	M192P47294
.0056	0.225	0.562	M192P56294
.0068	0.225	0.562	M192P68294
.0082	0.225	0.562	M192P82294
.01	0.225	0.562	M192P10394
.012	0.290	0.500	M192P12394
.015	0.290	0.500	M192P15394
.018	0.290	0.625	M192P18394
.022	0.290	0.625	M192P22394
.027	0.318	0.625	M192P27394
.033	0.318	0.625	M192P33394
.039	0.318	0.875	M192P39394
.047	0.318	0.875	M192P47394
.056	0.318	1.187	M192P56394
.068	0.318	1.187	M192P68394

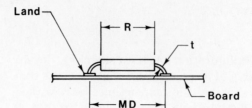

FIGURE 6-17 How to determine the distance between resistor mounting holes.

a resistor is mounted using wires that protrude from each side, which means that mounting size is larger than the resistor size.

The distance between the holes used to mount resistors (and cylindrically shaped capacitors and diodes) may be determined by the formula (see Figure 6-17)

$$MD = R + 8t$$

where MD = mounting distance
   R = component length
   t = wire thickness

However, almost all components have some "play" in their mounting distances. This means that the distances can be varied — made a little longer or shorter — as required. Mounting distances are usually matched to a background grid for manufacturing considerations. If, for example, the mounting distance were calculated to be 0.72 inch, an eight-to-the-inch pattern, the centers of the mounting holes would be located 0.75 inch apart. On a ten-to-the-inch pattern, it would be 0.70 inch. Figure 6-18 shows this concept.

FIGURE 6-18 Mounting distances should align with a grid background.

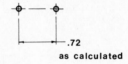

.72
as calculated

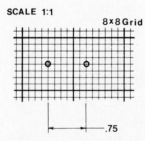

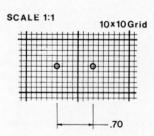

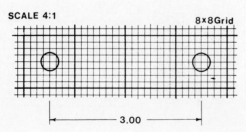

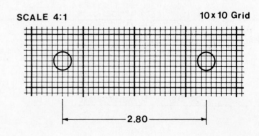

Eight-to-the-inch, ten-to-the-inch, and metric patterns are the most popular. Ten-to-the-inch patterns have become increasingly popular, as IC chips have mounting pins spaced 0.10 inch apart. The distance 0.10 inch is almost equal to 2.5 mm, so chips made to this mounting distance can easily be understood by designers and manufacturers who work in the metric system (see Figure 6-19).

Space allocation layouts are usually prepared larger than actual size. This is referred to as drawing "up". A computer's ZOOM function enables the designer to work at any given scale, although a final scale of 4:1 is most popular.

An increased scale is used to help ensure drawing, and therefore manufacturing accuracy. Any errors in sizing will be reduced when the drawing is reduced to actual size.

After a grid pattern and scale have been chosen and affixed to the drawing board, the next step is to define the component border, the PC board size, and the grid locators, as shown in Figure 6-20.

The *component border* is the area of the PC board that can be used to mount components. This area differs from the overall PC board area because it does not include any areas needed for mounting the board or which would interfere with other devices or boards. The component border is clearly defined by a heavy black line.

The *PC board size* is defined by using corner brackets. The corner brackets are heavy black lines drawn perpendicular to each other and to each corner of the board. The inside of the bracket should match the outside of the board. A single linear dimension is included on the layout together with the note "REDUCE TO . . ." to define the final board size. A note such as "REDUCE TO 2.00 ± .01" would mean that during the

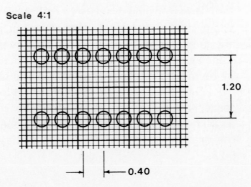

FIGURE 6-19 Layout should match the grid background if possible.

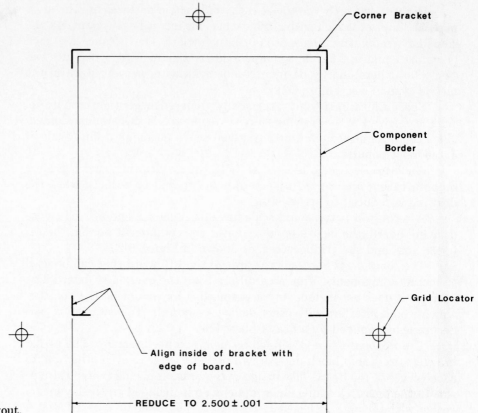

FIGURE 6-20   PC board layout.

final photographic processes in manufacturing, this dimension would be reduced until it exactly matched the specified 2.00 ± .01 inches.

Grid locators are symbols placed on the drawing to identify the location of the grid and to be used to help align the space allocation layout with related PC drawings. The crossed perpendicular lines must align exactly with the lines of the grid pattern. The circle is approximately 0.25 inch in diameter.

Component sizes and shapes will vary according to their manufacturer and performance capabilities (see Figure 6-16). Designers usually create their own component file based on the components they use most often. The file is called a component library and is similar to the electronic symbol library described in Section 4-2. Appendix H of this book includes some standard component sizes that could serve as a foundation for a component library.

It is common practice to call up all components needed for a drawing at one time and to locate them randomly around the edge of the PC drawing as shown in Figure 6-21. This saves going back to the file, which in some systems is time consuming. When needed, the components can easily be moved into position within the working area of the drawing.

Grid backgrounds are helpful in locating and aligning components, corner brackets, the component boarder, and grid indicators. Figure 6-22 shows a component drawing done using a grid background. After the drawing is complete, the grid may be erased as shown in Figure 6-23.

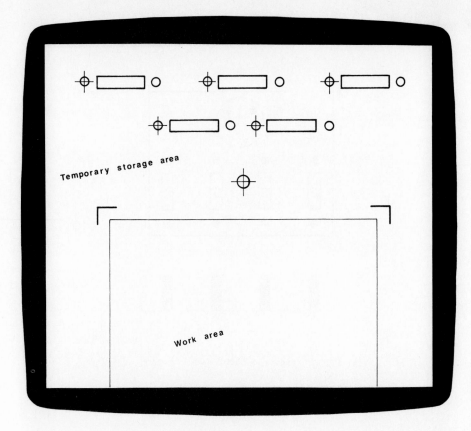

FIGURE 6-21 Call up all symbols needed for a drawing and locate them around the work area until needed.

FIGURE 6-22 Components located on a PC board design.

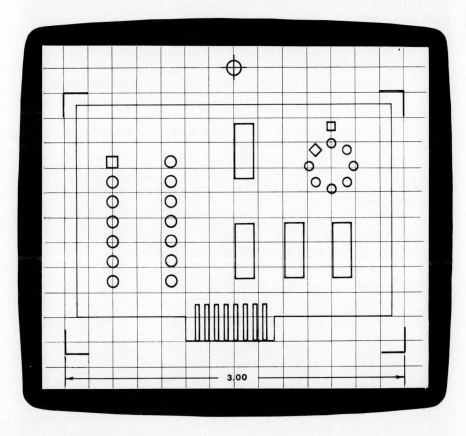

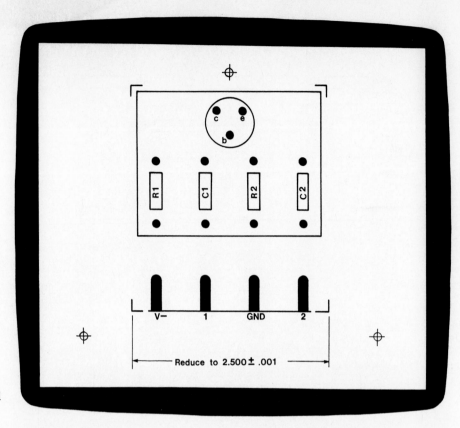

FIGURE 6-23 A completed
PC board drawing.

Figure 6-23 shows a space allocation layout for the schematic diagram presented in Figures 6-2 and 6-10. The drawing was prepared using a scale of 4:1. This means that the resistors, which actually measure 0.30 inch long, were drawn 1.20 inches long. All other components were drawn in a similar manner.

The components are always located on the top of the board; that is, the layout looks directly down on the components. The conductor paths are on the other side (the side not directly seen) of the board. Figure 6-24 shows this concept. Figure 6-25 shows the component side of a PC board. Figure 6-26 shows the conductor path side.

FIGURE 6-24 Layouts are
drawn as if viewed from the
component side of the board.

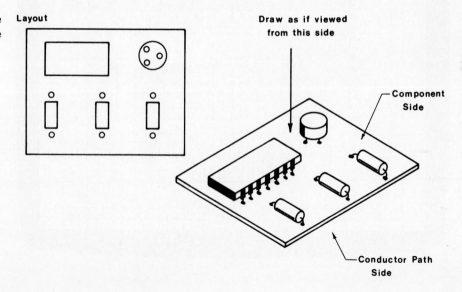

FIGURE 6-25   The component side of a PC board.

FIGURE 6-26   The conductor path side of a PC board.

Figures 6-27 and 6-28 are further examples of space allocation layouts based on the schematic diagrams shown in Figures 6-3 and 6-4. They have also been drawn at a scale of 4:1.

6-5 CONDUCTOR PATHS

Conductor paths are added to space allocations using solid lines, as shown in Figure 6-29. The paths are usually drawn using a series of straight lines (not curved lines), as shown.

Conductor paths are usually $\frac{1}{16}$ (0.0625) inch or approximately 1.5 mm wide with a minimum of $\frac{1}{32}$ (0.031) inch or 0.75 mm between the paths, as shown in Figure 6-30. For very lower power $\frac{1}{32}$-inch or 0.75 mm widths may be used and widths greater than $\frac{1}{16}$ inch or 1.5 mm are used for large power requirements.

The correct line spacing for conductor paths on a 4:1 scaled drawing

FIGURE 6-27 Space allocation layout.

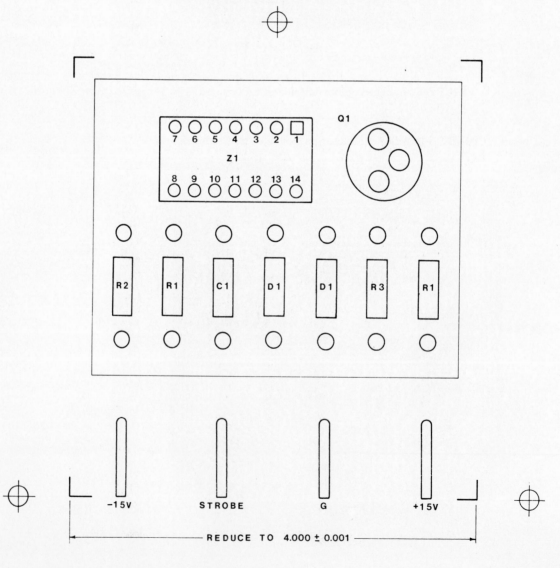

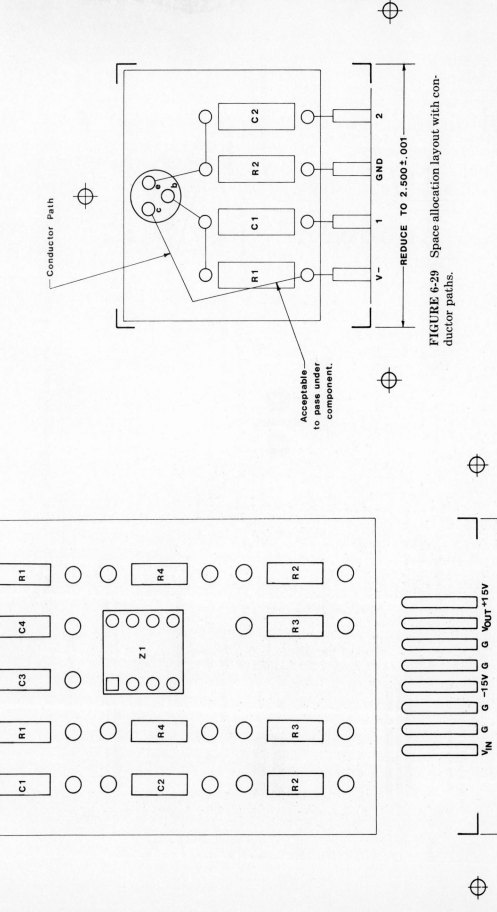

FIGURE 6-29 Space allocation layout with conductor paths.

FIGURE 6-28 Space allocation layout.

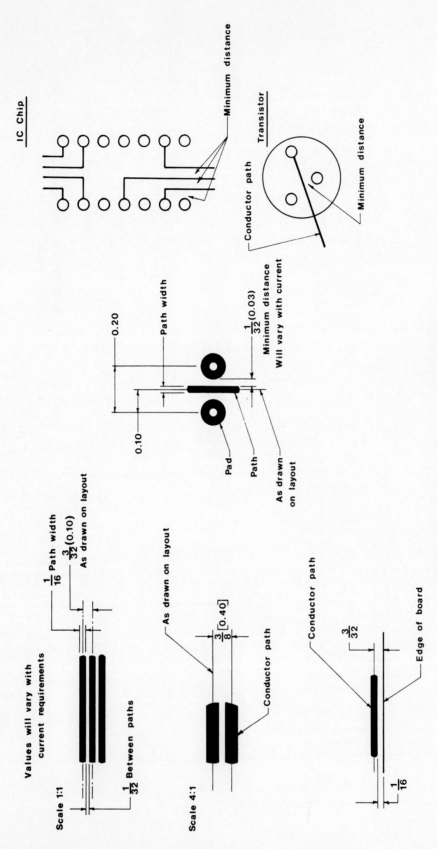

FIGURE 6-30 Layout requirements for conductor paths.

114

is found by the formula

$$D = 4W \times 4S$$

where D = distance between lines
$\quad W$ = conductor path width
$\quad S$ = width of space between conductor paths

This means that if the conductor path is $\frac{1}{16}$ inch wide and the spacing between paths is $\frac{1}{32}$ inch:

$$D = 4(\tfrac{1}{16}) + 4(\tfrac{1}{32})$$

$$= \tfrac{1}{4} + \tfrac{1}{8}$$

$$= \tfrac{3}{8} \text{ inch}$$

This may be expressed metrically as

$$D = 4(1.5) + 4(0.75)$$

$$= 6.00 + 3.00$$

$$= 9.00 \text{ mm}$$

Thus all lines that represent conductor paths should be drawn at least $\frac{3}{8}$ inch or 9 mm apart (see Figure 6-31).

Conductor paths may pass under IC chips, depending on the size of the chip and the conductor path width (see Figure 6-32).

Figures 6-33 and 6-34 are completed space allocation layouts with conductor paths added for the schematics presented in Figures 6-3 and 6-4.

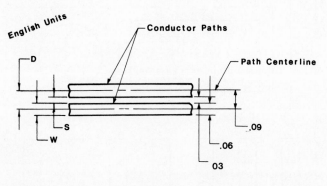

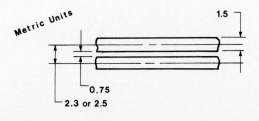

FIGURE 6-31 Spacing for conductor paths.

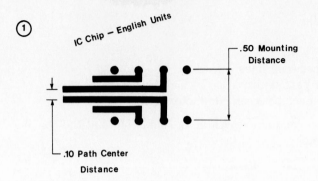

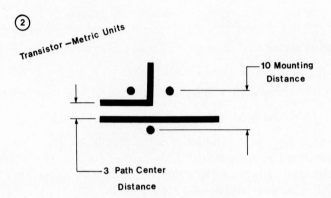

FIGURE 6-32 How to locate and space conductor paths under integrated circuits.

FIGURE 6-33 Sample space allocation layout based on the schematic shown in Figure 6-3.

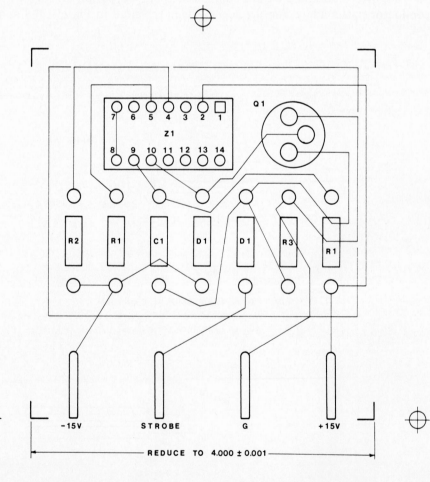

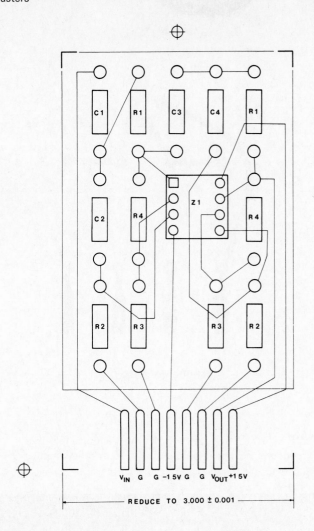

FIGURE 6-34 Sample space allocation layout based on the schematic shown in Figure 6-4.

6-6 FILM MASTERS

Film masters are special drawings produced on drawing film. Drawing film is similar to drawing paper but is thicker and much more durable. Film masters are used in the manufacturing of PC boards for both silk-screening and photographic processes.

Film masters are prepared on a computer by working with space allocation drawings. The final film master contains only the conductor paths, at their correct width and spacing, and the connector pads as shown in Figure 6-35. Corner brackets, grid locators, and a final reference dimension are also included. As with other PC drawings, film masters are prepared at a scale larger than actual size. A scale of 4:1 is common.

Figure 6-36 shows another example of a film master drawing. Figures 6-37 and 6-38 are tape masters for the schematics presented in Figures 6-3 and 6-4.

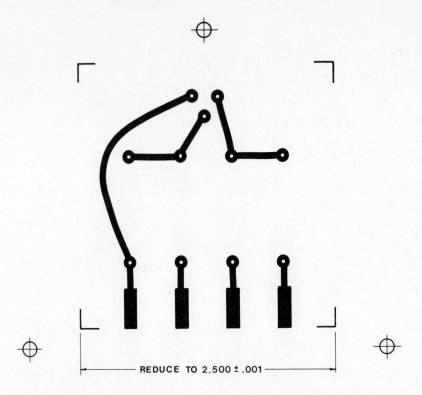

REDUCE TO 2.500 ± .001

FIGURE 6-35 Film master.

FIGURE 6-36 EE Designer ™ II from Visionics features surface-mount design capability. (Courtesy of Visionics Corp., Sunnyvale, Calif.)

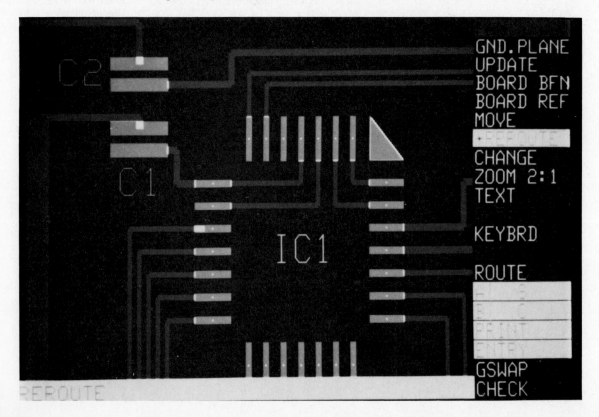

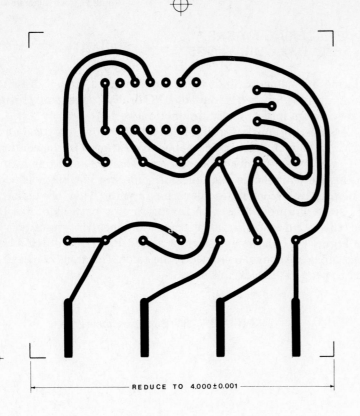

- REDUCE TO 4.000±0.001 -

FIGURE 6-37 Tape master for the schematic shown in Figure 6-3.

FIGURE 6-38 Tape master for the schematic shown in Figure 6-4.

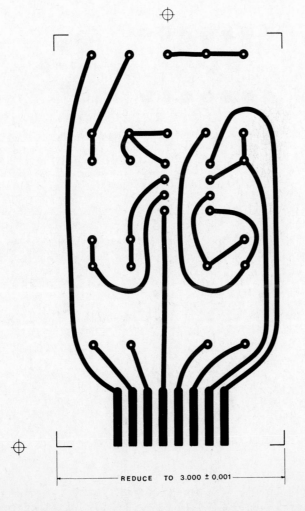

- REDUCE TO 3.000 ± 0.001 -

119

6-7   SOLDERING MASKS
AND DRILL DRAWINGS

Soldering masks and drill drawings are special types of drawings which
are not always prepared as part of a PC drawing package. However, they
are very useful in automatic production work.

A soldering mask is a film master that shows only the pad areas of
the circuit. It includes a solid shape (no clear area in the center) for each
pad as necessary, corner brackets, grid locators, and one sizing dimension. Figure 6-39 shows examples of soldering masks. Soldering masks
can be prepared from space allocation layouts. They are usually done at
a 4:1 scale. Figure 6-40 shows the conductor path side of a PC board.
Note the round solder areas at each component intersection.

Figure 6-41 shows a design for a multilayer board. Note how all soldering areas are shown using the same color. The areas can easily be identified to prepare a soldering mask.

FIGURE 6-39   Soldering master.

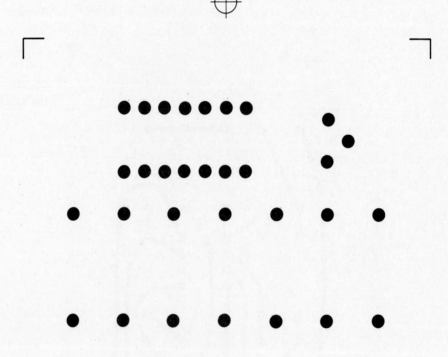

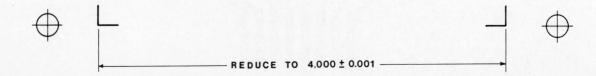

REDUCE TO   4.000 ± 0.001

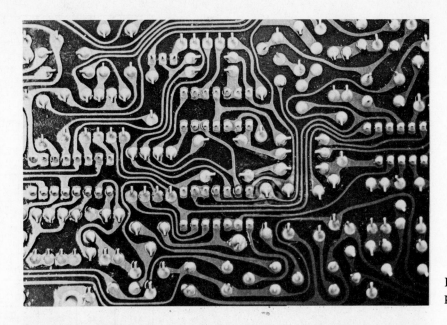

FIGURE 6-40 The conductor path side of a PC board.

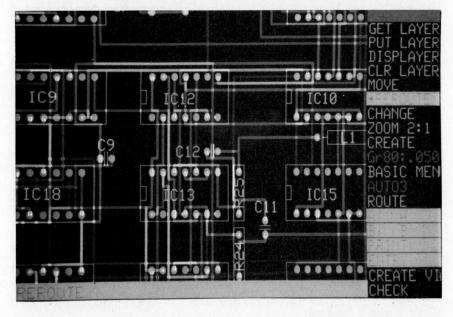

FIGURE 6-41 EE Designer™ II from Visionics features multi-layer capability. (Courtesy of Visionics Corp., Sunnyvale, Calif.)

Drill drawings are drawings that define the pattern of holes which must be drilled into the board using an x-y coordinate system. Figure 6-42 shows an example.

The hole pattern is traced from the space allocation layout. A zero reference point is chosen, which may be the corner of the board or the center of a locating hole previously drilled into the board. The x and y directions are defined from the zero reference point, and the minimum number of dimensions needed to size the board are added.

All hole dimensions are defined in terms of the x-y coordinate system and presented in a chart form as shown. Note that the drawing is a trace of a drawing done at 4:1 scale; *all dimensions listed in the chart must be actual size.* The chart dimensions are final dimensions, that is, the size to be used in manufacturing. Divide the traced dimensions by 4 to get the actual dimensions.

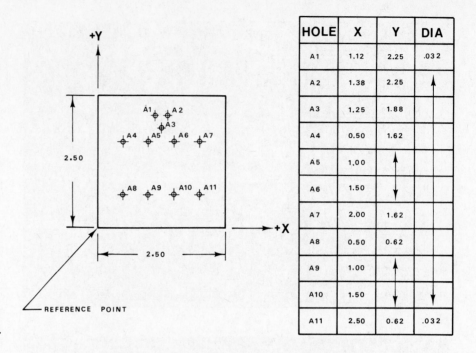

HOLE	X	Y	DIA
A1	1.12	2.25	.032
A2	1.38	2.25	↑
A3	1.25	1.88	
A4	0.50	1.62	
A5	1.00	↑	
A6	1.50	↓	
A7	2.00	1.62	
A8	0.50	0.62	
A9	1.00	↑	
A10	1.50	↓	↓
A11	2.50	0.62	.032

FIGURE 6-42   Drill drawing.

To prevent error, some drafters prefer to redraw the hole pattern at a scale of 1:1 so that measurements may be taken directly. However, it is a fairly simple process to divide all measured values by 4, especially if a calculator is available.

6-8   DOUBLE-SIDED PC BOARDS

Double-sided PC boards differ from single-sided boards in that they have conductor paths on boths sides of the board. Components are still usually restricted to one side of the board but could be mounted on both sides. Double-sided boards are more expensive and time consuming to manufacture but use a smaller area than that of single-sided boards. Figure 6-43 shows the component side of a double-sided PC board. Note that it contains both components and conductor paths. Figure 6-44 shows the conductor path side of the board. Components could be mounted to this side of the board if necessary.

Drawings for double-sided boards are prepared using the same procedure as that used for single-sided boards. There is one space allocation layout (master layout), but film masters must be prepared for each side of the board. The increased number of drawings requires:

1.   Accurate grid locators
2.   Careful identification of each drawing

It is vital that the individual drawings match each other. Any drawing errors will result in manufacturing errors.

Figure 6-45 shows a space allocation layout developed from a schematic. The procedure used to locate the components and conductor paths is as explained previously, but some differences are required to distinguish which paths are on which side of the board.

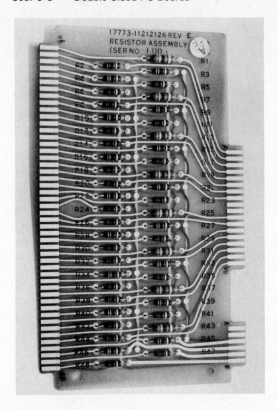

FIGURE 6-43 The component side of a double-sided PC board.

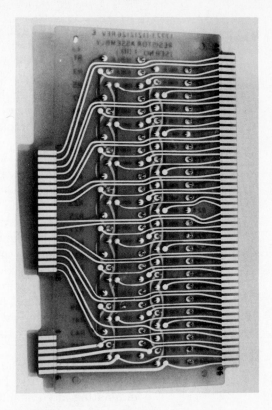

FIGURE 6-44 The conductor path side of a double-sided PC board.

FIGURE 6-45 Space allocation drawing for a double-sided PC board.

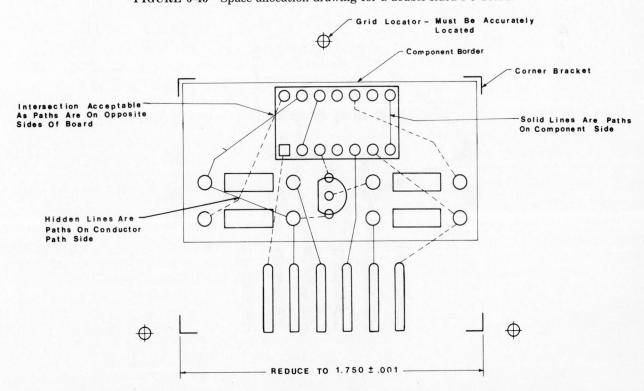

The board is still viewed from the component side. The component side is considered to be the top of the board. Conductor paths on the top side (component side) use a solid line; conductor paths on the bottom side (conductor path side) use hidden lines. Computer systems use different color lines to distinguish board side. Figure 6-46 shows an example of how different color paths may be used. The hidden line is used for clarity. Film masters are prepared for each side of the board using the common master layout.

The component side is prepared using red lines, the conductor path side, using blue lines. Both sides of the drawing film are used, so when viewed from one angle the lettering on the opposite side will be reversed. During manufacturing, special camera filters are used to block out the unwanted color pattern.

Another technique, shown in Figure 6-47, uses two separate film masters. One defines the conductor path side pattern, the other, the component side pattern. Sometimes the lettering on the conductor path is done backward to help call attention to the fact that it is opposite the conductor path side.

A third technique, shown in Figure 6-48, uses one large sheet of drawing film. Both sides are drawn on the same sheet, separated by an end view.

As with single-sided boards, pad masters are sometimes prepared for production work. Pad masters ensure more accurate production by eliminating any mismatch between pads drawn on opposite sides of the board.

FIGURE 6-46   Identifying a double-sided PC board.

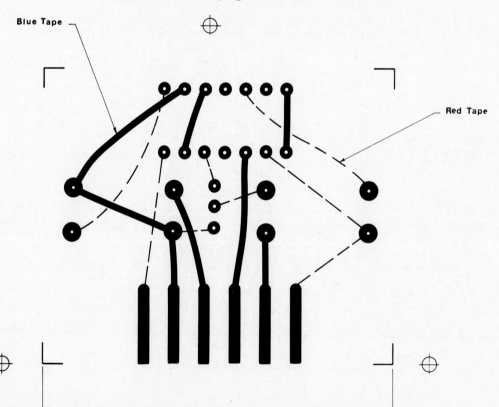

Component Side

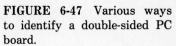

FIGURE 6-47 Various ways to identify a double-sided PC board.

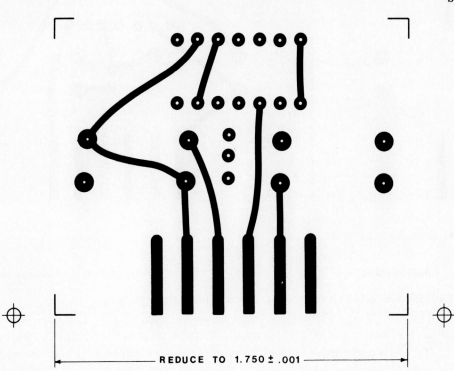

REDUCE TO 1.750 ± .001

COMPONENT SIDE

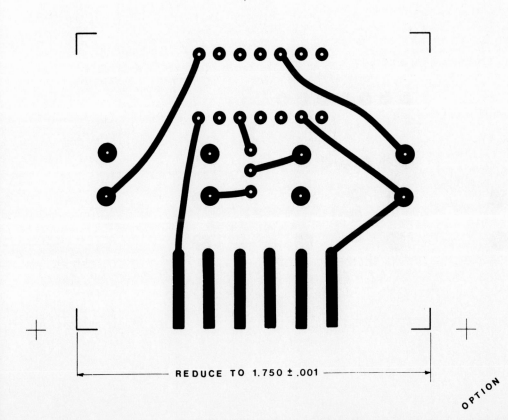

REDUCE TO 1.750 ± .001

OPTION

CONDUCTOR PATH SIDE

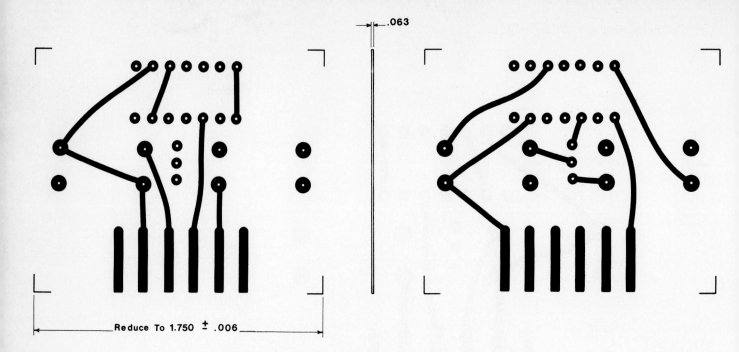

.063

Reduce To 1.750 ± .006

Component Side

Conductor Path Side

FIGURE 6-48 Drawing that includes both sides of a double-sided PC board.

If a pad master is used, three film masters will be required: the pad master, the conductor paths on the component side, and the conductor paths on the conductor side. Figure 6-49 shows a set of film masters that includes a pad master. Conductor paths are always overlapped onto the pad areas.

FIGURE 6-49(a) Pad master.

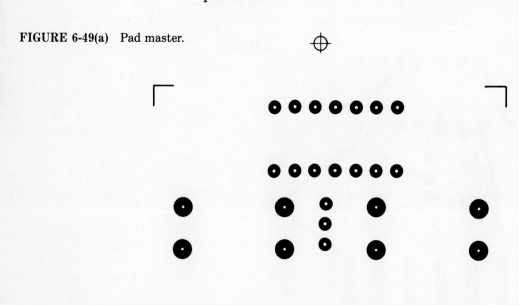

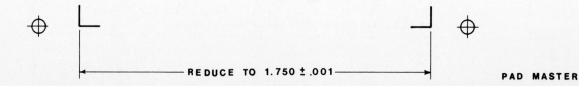

REDUCE TO 1.750 ± .001

PAD MASTER

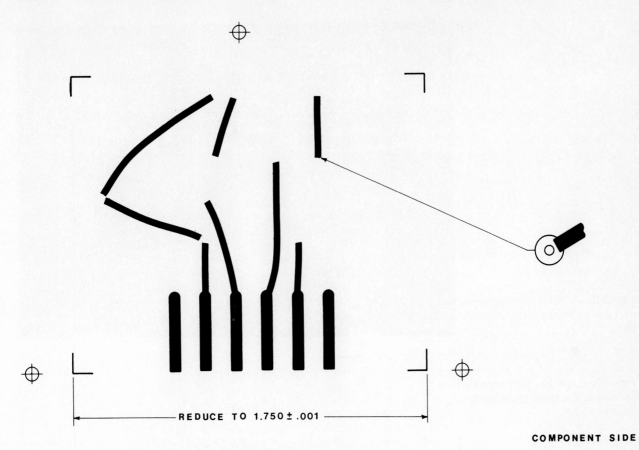

REDUCE TO 1.750 ± .001

COMPONENT SIDE

(b) Component-side tape drawing.

(c) Conductor-path-side tape drawing.

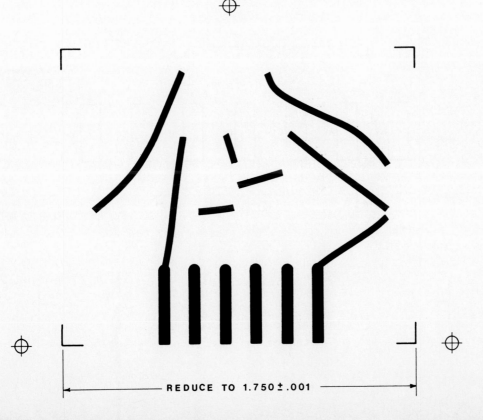

REDUCE TO 1.750 ± .001   CONDUCTOR PATH SIDE

FIGURE   6-50 Component outlines and identification numbers.

FIGURE   6-51 Component outlines and identification number drawing.

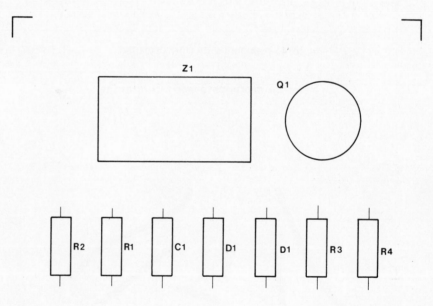

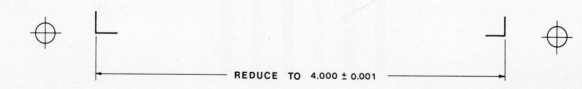

REDUCE TO 4.000 ± 0.001

6-9   COMPONENT OUTLINE DRAWINGS

Component outline drawings are drawings that outline the overall component shape and add identification numbers to the component surface of a board. Figure 6-50 shows a photo of a board with component outlines and identification numbers. Component outlines and identification numbers make it easier to assemble the board and easier to identify components during board maintenance or troubleshooting.

Component outline drawings are prepared from the master layout. Figure 6-51 shows a sample component outline drawing. Note that the identification numbers are placed next to the appropriate component so that they can still be seen on the board after the component has been mounted.

The outline sizes are drawn slightly larger than the actual component. The outline sizes may be varied as needed provided that they *clearly* distinguish the component.

Problems   6-1 thru 6-5

P6-1 through P6-9   Based on the schematic diagrams in Figures P6-1 through P6-9, prepare the following types of drawings.

SKETCH

(a) A single-sided space allocation drawing
(b) A double-sided space allocation drawing
(c) A tape master
(d) Soldering masks

DRAWINGS

(e) A component outline drawing

Use the component sizes as defined in Appendix H.

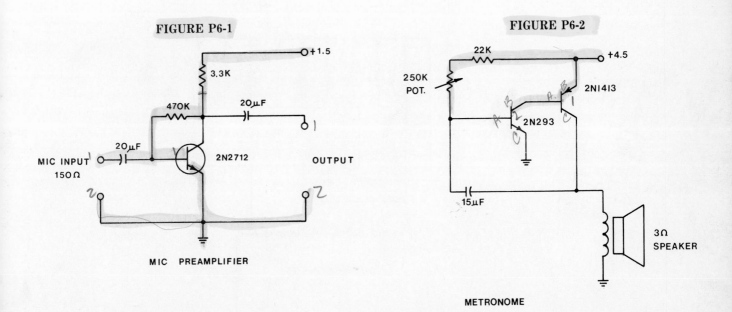

FIGURE P6-1

3.3K
470K   20μF
+1.5
20μF   2N2712   OUTPUT
MIC INPUT
150Ω

MIC   PREAMPLIFIER

FIGURE P6-2

22K   +4.5
250K
POT.   2N1413
2N293
15μF
3Ω
SPEAKER

METRONOME

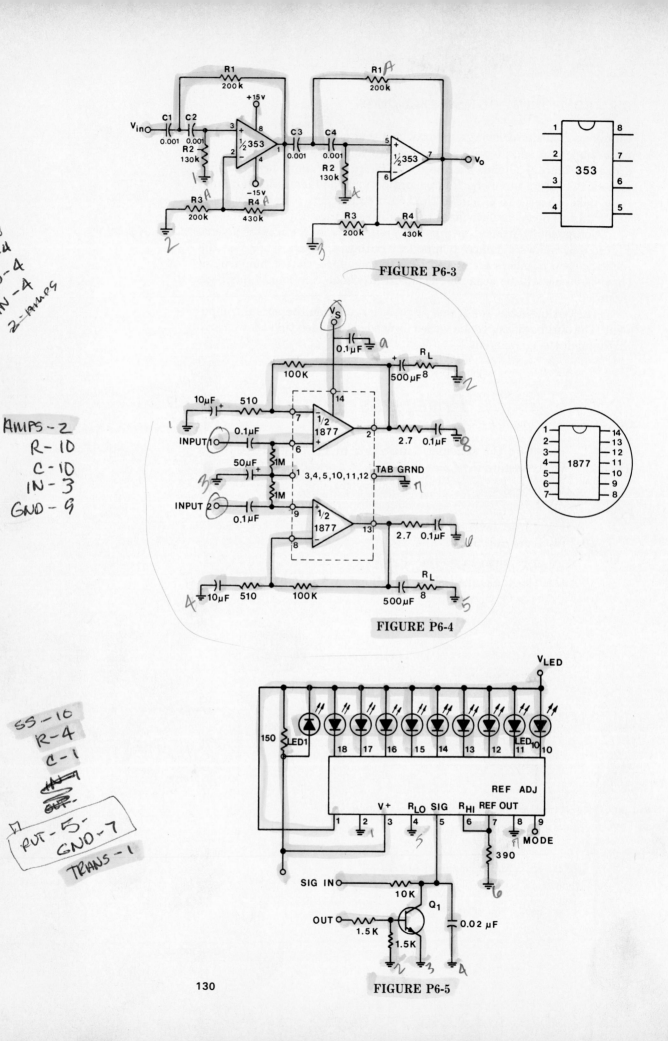

FIGURE P6-3

FIGURE P6-4

FIGURE P6-5

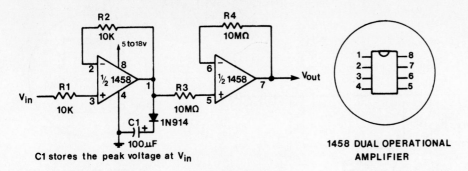

C1 stores the peak voltage at V$_{in}$

1458 DUAL OPERATIONAL
AMPLIFIER

FIGURE P6-6

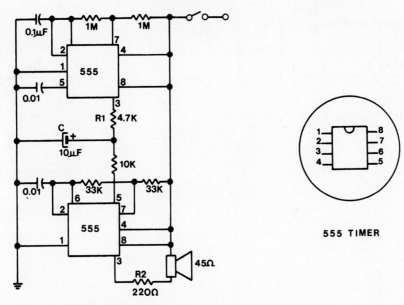

Warble Alarm Circuit

555 TIMER

FIGURE P6-7

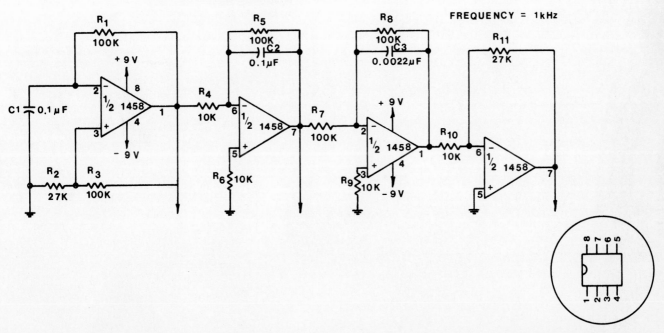

FREQUENCY = 1kHz

1458 DUAL OPERATIONAL
AMPLIFIER

FIGURE P6-8

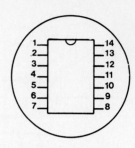

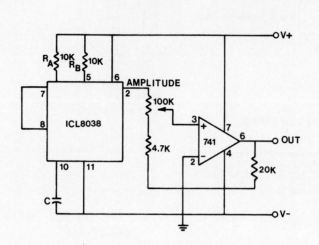

**ICL8038 PRECISION
WAVEFORM GENERATOR**

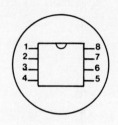

741 OPERATIONAL AMP

FIGURE P6-9

7

Block
and Logic Diagrams

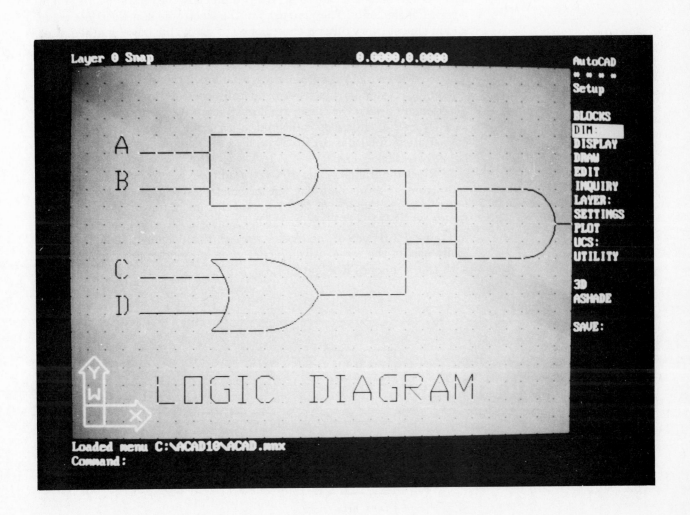

7-1 INTRODUCTION

In this chapter we explain how to draw and interpret block and logic diagrams. As in previous chapters, the drawing fundamentals required are explained in detail, but in addition, a discussion of the use of each type of diagram is also included.

7-2 BLOCK DIAGRAMS

Block diagrams are a way to express graphically the relationships among a series of elements. They are used in the electronic and electrical fields, and in almost every other technical area of study: business, mathematics, engineering, to name just a few. Figure 7-1 is an example of a block diagram that illustrates a stereo system.

To draw a block diagram, no special block size or shape is required, although rectangular blocks are generally used. Usually, all blocks on a diagram are drawn the same size and all lines are drawn the same thickness. If special emphasis is desired, a larger block or a thicker line or a combination of both is drawn. Note how the amp in Figure 7-1 is emphasized by a thicker line.

Block diagrams are set up to be read from left to right and from top to bottom. If more than one line of blocks is required, the second line must be located under the first row and the line of flow must be returned from the end of the first row to the beginning of the second row, as shown in Figure 7-2. If one of the blocks represents an element that is not normally part of the series, it can be drawn using hidden lines, as was done with the earphones in Figure 7-1.

FIGURE 7-1 Block diagram of a stereo system.

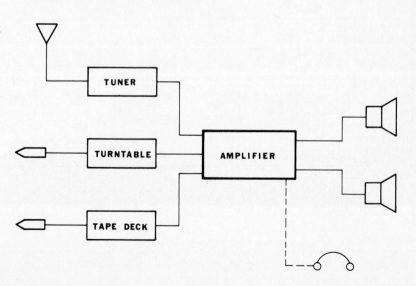

134

As there are no special sizes for block diagrams, they are usually drawn by aligning the figures with a grid background as shown in Figure 7-3.

Block diagrams are used not only to express graphically the relationship between a series of elements but can also be used to analyze a series of elements. Consider the following problem. You have a coin, one side of which is a head and the other side a tail, and you are going to flip the coin four times. What are the chances of getting a head four consecu-

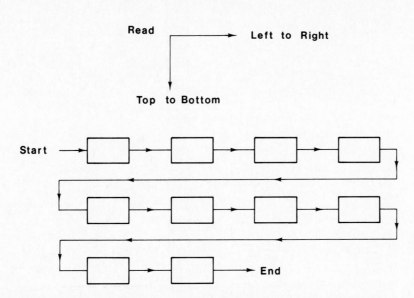

FIGURE 7-2 Block diagrams are read from left to right and top to bottom.

FIGURE 7-3 Grids are used to both position and size the rectangles used in block diagrams.

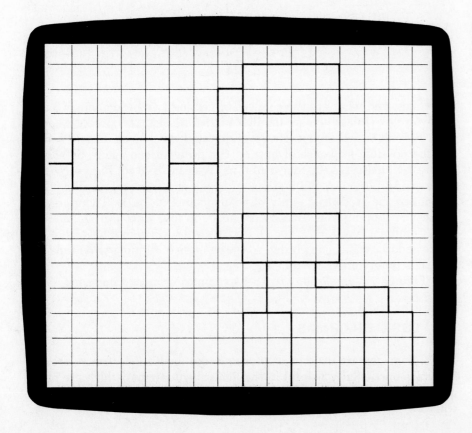

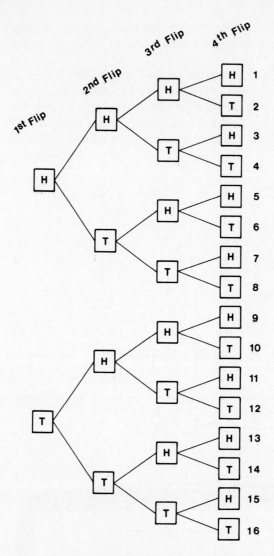

FIGURE 7-4 Block diagram that illustrates four flips of a coin.

tive times, and what are the chances of getting two heads and two tails in any order? Unless you are familiar with statistics, this problem can be very difficult to reason out. However, if you draw a block diagram showing the possible sequence of events, the problem can easily be solved. Figure 7-4 illustrates the four coin flips. We can see from Figure 7-4 that there are 16 possible combinations of heads and tails in four flips. Only one out of the 16 consists only of heads (combination 1). Therefore, the chances of flipping four consecutive heads in four flips are 1 in 16.

Combinations 4, 6, 7, 10, 11, and 13 will yield combinations of two heads and two tails in any order, so the chances of flipping two heads and two tails in any order are 6 in 16.

Figures 7-5 and 7-6 are further examples of block diagrams.

7-3 LOGIC DIAGRAMS

Logic diagrams are used to help design and analyze electronic circuits. They are particularly helpful in the analysis of circuits which are based on binary principles, such as those found in computers and electronic calculators.

FILTER   CIRCUIT

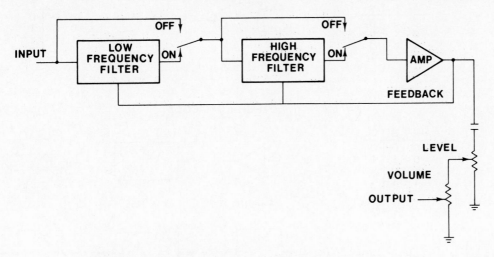

FIGURE 7-5   Example of a block diagram. (Courtesy of General Electric Company.)

RECEIVER

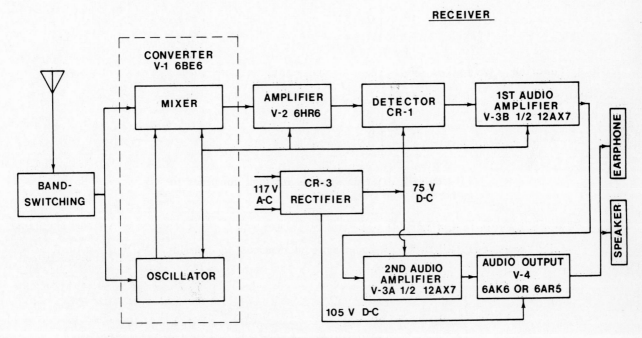

FIGURE 7-6   Example of a block diagram. (Courtesy of General Electric Company.)

There are several different logic functions, each of which has its own logic symbol. The logic symbols are defined in Figure 7-7 and the dimensions for some of the symbols are presented in Figure 7-8. The sizes defined in Figure 7-8 are in concurrence with the Department of Defense standard MIL-STD-806, Graphic Symbols for Logic Diagrams. These sizes are considered national standards, but may be varied as long as the basic shape remains the same.

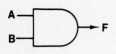

AND   Function

A	B	F
0	0	0
0	1	0
1	0	0
1	1	1

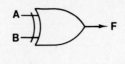

EXCLUSIVELY OR   Function

A	B	F
0	0	0
0	1	1
1	0	1
1	1	0

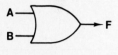

OR   Function

A	B	F
0	0	0
0	1	1
1	0	1
1	1	1

OTHER   SYMBOLS

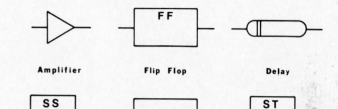

Amplifier   Flip Flop   Delay

Single Shot   General   Schmitt Trigger

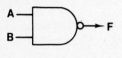

NAND   Function   (negative AND)

A	B	F
0	0	1
0	1	1
1	0	1
1	1	0

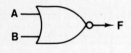

NOR   Function   (negative OR)

A	B	F
0	0	1
0	1	0
1	0	0
1	1	0

FIGURE 7-7 Graphic symbols for logic functions together with their truth tables.

FIGURE 7-8 Recommended dimensions for various logic symbols.

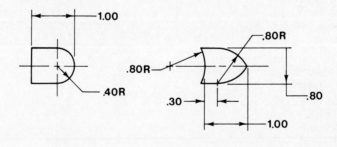

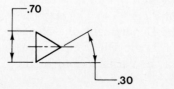

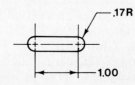

Designers file logic symbols in a logic symbol library. Figure 7-9 shows a possible logic symbol library.

Logic diagrams are analyzed using *truth tables*. Figure 7-7 shows several basic logic functions and their truth tables. Figure 7-10 shows an AND gate symbol and a truth table four an AND gate. An AND gate passes current only when both inputs A and B are active. The inputs A and B generate four possible combinations as shown on the left side of the truth table. Figure 7-10 further shows some simple circuits that help explain how an AND gate reacts to inputs. If inputs A and B were off–on switches as shown, current would pass through the circuit only when both the A switch and the B switch were closed — thus the name AND gate.

Figure 7-11 shows an OR gate with its truth table and an appropriate simplified circuit. OR gates will pass current if either the A switch or the B switch is closed — thus the name OR gate.

Truth tables are used to analyze not only individual logic functions but also those used in combination. For example, study the circuit shown in Figure 7-12. If the input is $A = 1$, $B = 0$, $C = 0$, the AND function will generate 0 output. When this 0 output is combined with the $C = 0$ input into the OR function, the final output will be 0. If we change the inputs to $A = 0$, $B = 0$, $C = 1$, the AND function will produce a 0 output. This 0 coupled with the $C = 1$ input will, when acted upon by the OR function, generate a final output of 1. The truth table shown in Figure 7-12 analyzes all the possible combinations for the circuit.

Sometimes it is difficult to remember all the possible combinations that a number of inputs could generate. Figure 7-13 has been included to help define all the possible combinations for up to five given inputs.

FIGURE 7-9 An example of a Logic Symbols file.

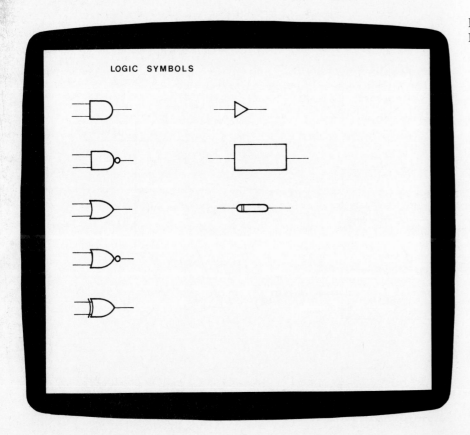

LOGIC SYMBOLS

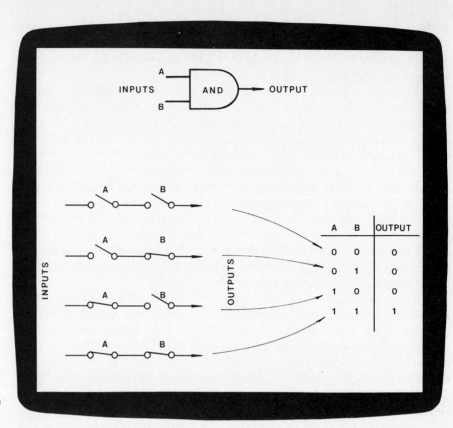

FIGURE 7-10 An AND gate and its truth table.

FIGURE 7-11 An OR gate and its truth table.

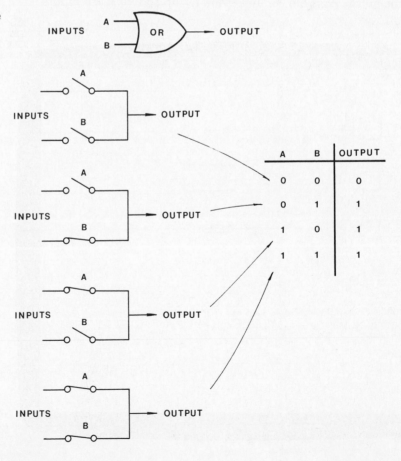

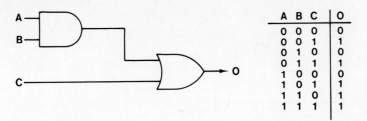

A	B	C	O
0	0	0	0
0	0	1	1
0	1	0	0
0	1	1	1
1	0	0	0
1	0	1	1
1	1	0	1
1	1	1	1

FIGURE 7-12 Example of a logic diagram together with its truth table.

FIGURE 7-13 Possible combinations for up to five inputs. Note that for two inputs there are four possible combinations; for three inputs, eight combinations; for four inputs, 16 combinations; for five inputs, 32 combinations; and so on.

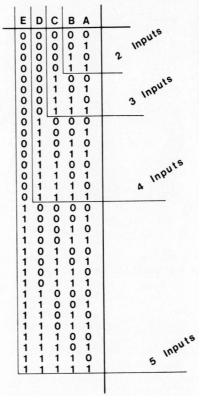

Figures 7-14 and 7-15 are examples of logic diagrams together with their truth tables. Study each one and verify the meaning of each set of input combinations.

Figure 7-16 compares some logic diagrams with their equivalent schematic representations.

The notations 0 and 1 as used in truth tables are not numerical digits. They do *not* mean zero current or a current of unit value 1. The notations 0 and 1 represent the on and off function of a gate. They can also represent low and high voltages within a circuit. Figure 7-17 shows three different types of truth tables for an AND gate. The three tables are equivalent; only the notations are different.

Truth tables may be shortened when certain inputs are obvious or when only a certain number of inputs are possible. For example, the truth table for the logic circuit shown in Figure 7-18 has a maximum of 32 possible outputs. But if we know that input A is always 1, and inputs D and E are always 0, we can rewrite the truth table as shown.

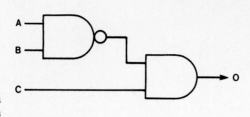

A	B	C	O
0	0	0	0
0	0	1	1
0	1	0	0
0	1	1	1
1	0	0	0
1	0	1	1
1	1	0	0
1	1	1	0

FIGURE 7-14 Example of a logic diagram together with its truth table.

FIGURE 7-15 Example of a logic diagram together with its truth table.

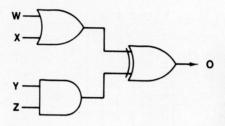

W	X	Y	Z	O
0	0	0	0	0
0	0	0	1	0
0	0	1	0	0
0	0	1	1	1
0	1	0	0	1
0	1	0	1	1
0	1	1	0	1
0	1	1	1	0
1	0	0	0	1
1	0	0	1	1
1	0	1	0	1
1	0	1	1	0
1	1	0	0	1
1	1	0	1	1
1	1	1	0	1
1	1	1	1	0

FIGURE 7-16 Some logic diagrams with their equivalent schematic diagrams.

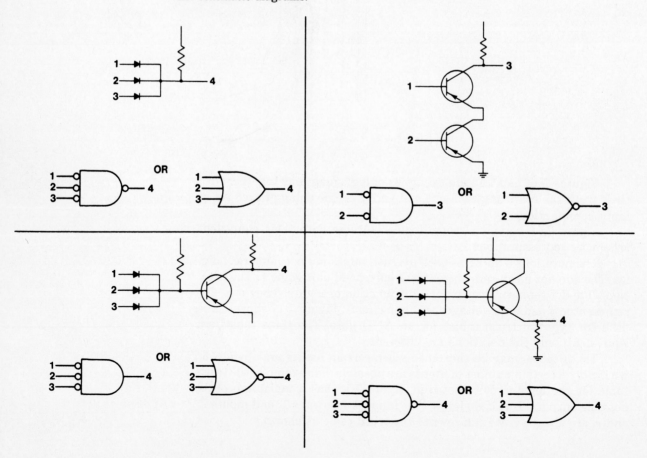

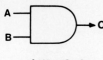

AND Gate

IN		OUT
A	B	C
OFF	OFF	OFF
OFF	ON	OFF
ON	OFF	ON
ON	ON	ON

IN		OUT
A	B	C
0	0	0
0	1	0
1	0	0
1	1	1

IN		OUT
A	B	C
L	L	L
L	H	L
H	L	L
H	H	H

FIGURE 7-17   Different types of truth-table notations.

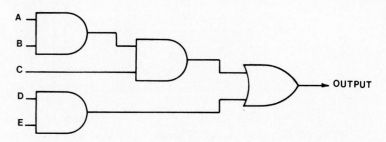

FIGURE 7-18   Shortened truth table.

E	D	C	B	A	OUTPUT
0	0	0	0	0	0
0	0	0	0	1	0
0	0	0	1	0	0
0	0	0	1	1	0
0	0	1	0	0	0
0	0	1	0	1	0
0	0	1	1	0	0
0	0	1	1	1	1
0	1	0	0	0	0
0	1	0	0	1	0
0	1	0	1	0	0
0	1	0	1	1	0
0	1	1	0	0	0
0	1	1	0	1	0
0	1	1	1	0	0
0	1	1	1	1	1
1	0	0	0	0	0
1	0	0	0	1	0
1	0	0	1	0	0
1	0	0	1	1	0
1	0	1	0	0	0
1	0	1	0	1	0
1	0	1	1	0	0
1	0	1	1	1	1
1	1	0	0	0	1
1	1	0	0	1	1
1	1	0	1	0	1
1	1	0	1	1	1
1	1	1	0	0	1
1	1	1	0	1	1
1	1	1	1	0	1
1	1	1	1	1	1

IF   A = 1   Reduce Table To
     D = 0
     E = 0

E	D	C	B	A	OUTPUT
0	0	0	0	1	0
0	0	0	1	1	0
0	0	1	0	1	0

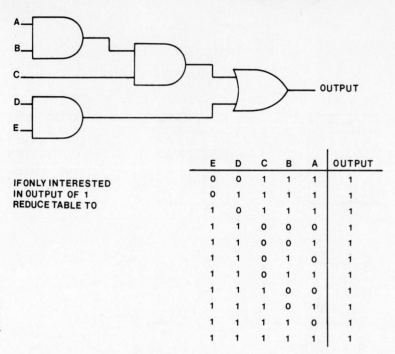

E	D	C	B	A	OUTPUT
0	0	1	1	1	1
0	1	1	1	1	1
1	0	1	1	1	1
1	1	0	0	0	1
1	1	0	0	1	1
1	1	0	1	0	1
1	1	0	1	1	1
1	1	1	0	0	1
1	1	1	0	1	1
1	1	1	1	0	1
1	1	1	1	1	1

IF ONLY INTERESTED
IN OUTPUT OF 1
REDUCE TABLE TO

FIGURE 7-19 Shortened truth table.

The truth table shown (Figure 7-19) also has 32 possible outputs. This time we are interested only in outputs that generate values of 1, so we can eliminate all combinations that yield 1 and shorten the table as shown.

7-4 LOGIC SYMBOLS AND SCHEMATIC DIAGRAMS

Logic symbols are often included as part of a schematic diagram to help simplify the diagrams. The logic symbols may be included as part of an IC to help define the IC's function in relation to the rest of the schematic, as shown in Figure 7-20, or it may be used as a separate function, as shown in Figure 7-21.

FIGURE 7-20 Logic symbols used to explain an IC's function within a circuit.

FIGURE 7-21 Logic symbols used to explain the function of an IC chip.

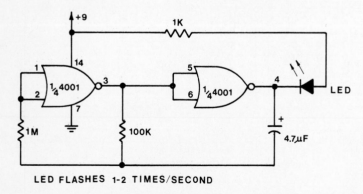

LED FLASHES 1-2 TIMES/SECOND

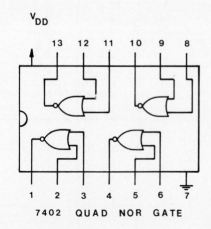

7402   QUAD   NOR   GATE

Logic symbols drawn in schematic diagrams are drawn using the size dimensions outlined in Figure 7-8. Schematics are still read from left to right, going from input to output.

Circuit manufacturers usually include both a truth table and a logic diagram as part of the circuit's specifications. This enables the designer to understand the circuit's function and thereby decide how it can be used in a design. Figure 7-22 shows a set of specifications from Texas Instruments for a data selector multiplexer.

FIGURE 7-22 A sample page from a Texas Instruments Catalog. Courtesy of Texas Instruments Inc.

TYPES SN5423, SN5425, SN7423, SN7425
DUAL 4-INPUT NOR GATES WITH STROBE

REVISED DECEMBER 1983

- **Package Options Include Plastic and Ceramic DIPs**
- **Dependable Texas Instruments Quality and Reliability**

description

These devices contain dual 4-input positive NOR gates with strobe. The SN5423 and SN7423 are expandable, and perform the Boolean functions:

$$1Y = \overline{1G(1A + 1B + 1C + 1D) + X} \text{ and}$$
$$2Y = \overline{2G(2A + 2B + 2C + 2D)}$$

with X = output of SN5460/SN7460. The SN5425 and SN7425 perform the Boolean function:

$$Y = \overline{G(A + B + C + D)}$$

The SN5423 and the SN5425 are characterized for operation over the full military temperature range of −55°C to 125°C. The SN7423 and the SN7425 are characterized for operation from 0°C to 70°C.

SN5423 . . . J OR W PACKAGE
SN7423 . . . J OR N PACKAGE
(TOP VIEW)

```
      ____
1X [ 1  U 16 ] VCC
1A [ 2    15 ] 1X
1B [ 3    14 ] 2D
1G [ 4    13 ] 2C
1C [ 5    12 ] 2G
1D [ 6    11 ] 2B
1Y [ 7    10 ] 2A
GND[ 8     9 ] 2Y
```

SN5425 . . . J OR W PACKAGE
SN7425 . . . J OR N PACKAGE
(TOP VIEW)

```
      ____
1A [ 1  U 14 ] VCC
1B [ 2    13 ] 2D
1G [ 3    12 ] 2C
1C [ 4    11 ] 2G
1D [ 5    10 ] 2B
1Y [ 6     9 ] 2A
GND[ 7     8 ] 2Y
```

schematic (each gate)

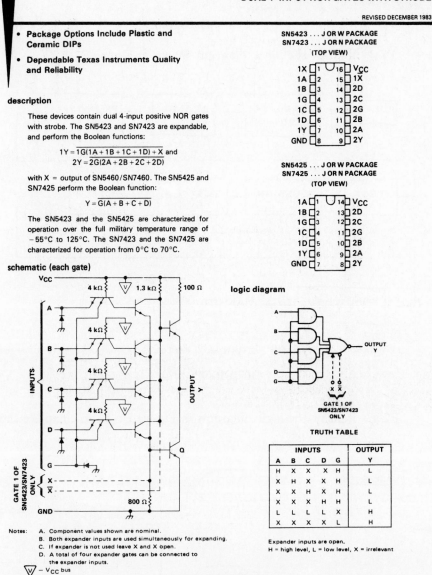

logic diagram

OUTPUT Y

GATE 1 OF
SN5423/SN7423
ONLY

TRUTH TABLE

INPUTS					OUTPUT
A	B	C	D	G	Y
H	X	X	X	H	L
X	H	X	X	H	L
X	X	H	X	H	L
X	X	X	H	H	L
L	L	L	L	X	H
X	X	X	X	L	H

Expander inputs are open.
H = high level, L = low level, X = irrelevant

Notes: A. Component values shown are nominal.
B. Both expander inputs are used simultaneously for expanding.
C. If expander is not used leave X and X open.
D. A total of four expander gates can be connected to the expander inputs.

 – VCC bus

TEXAS INSTRUMENTS
POST OFFICE BOX 225012 ● DALLAS, TEXAS 75265

Problems

P7-1   Redraw the block diagram shown in Figure 7-1 and add two more speakers.

P7-2   Prepare a block diagram that displays the following series of events.

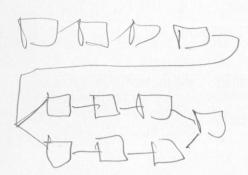

	Get up
	Go to school
	Go to class
	Go to lunch
Go to class	Play cards
Go home	Play more cards
Study	Go home
	Eat supper

P7-3   Redraw the block diagram shown in Figure P7-3.

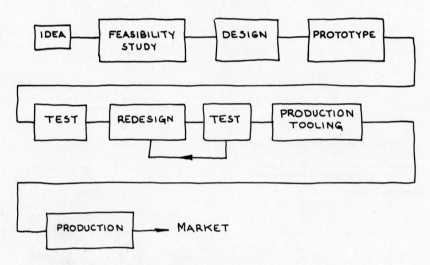

THE DESIGN PROCESS

FIGURE P7-3

P7-4   Redraw the logic diagrams in Figure P7-4 and complete the truth tables.

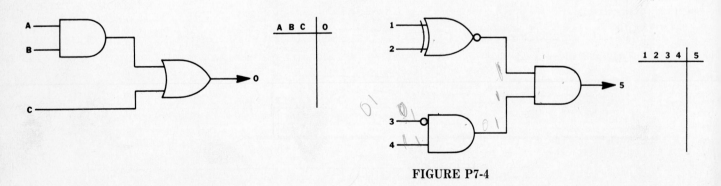

FIGURE P7-4

P7-5   Redraw the following truth table and add a logic diagram whose functions will generate the truth table.

A	B	C	Output
0	0	0	0
0	0	1	0
0	1	0	0
0	1	1	0
1	0	0	0
1	0	1	0
1	1	0	0
1	1	1	1

P7-6   Redraw the logic diagram shown in Figure P7-6 and add the truth table.

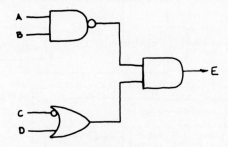

FIGURE P7-6

P7-7   Redraw the logic diagrams shown in Figure P7-7 and add the truth table.

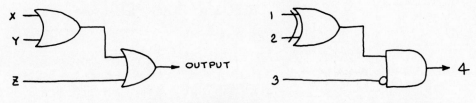

FIGURE P7-7

P7-8   Redraw the logic diagram shown in Figure P7-8 and complete a truth table for any 6 of the 64 possible input combinations.

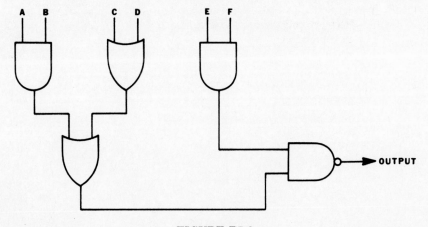

FIGURE P7-8

P7-9   Redraw Figures P7-9 through P7-12.
through
P7-12

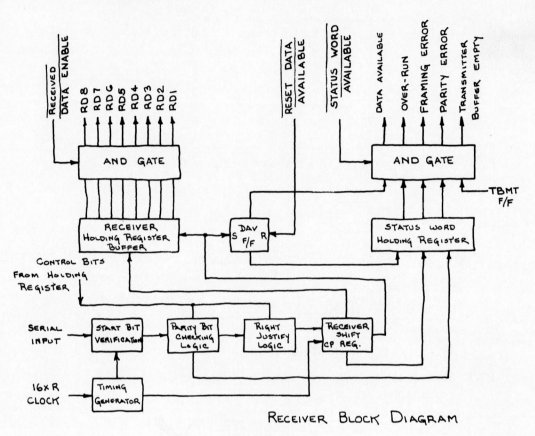

RECEIVER BLOCK DIAGRAM

FIGURE P7-9

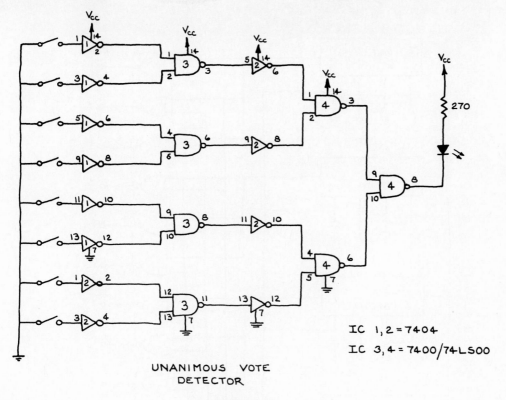

UNANIMOUS VOTE
DETECTOR

IC 1,2 = 7404
IC 3,4 = 7400/74LS00

FIGURE P7-10

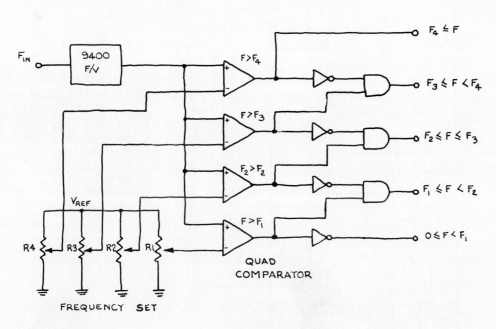

FREQUENCY/TONE DECODER

FIGURE P7-11

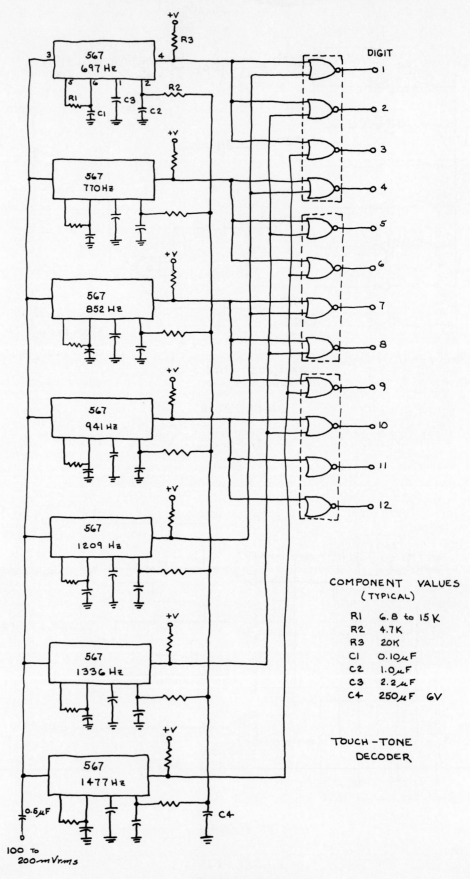

FIGURE P7-12

P7-13   Redraw the logic diagrams and corresponding equivalent cir-
and **P7-14**   cuits shown in Figures P7-13 and P7-14.

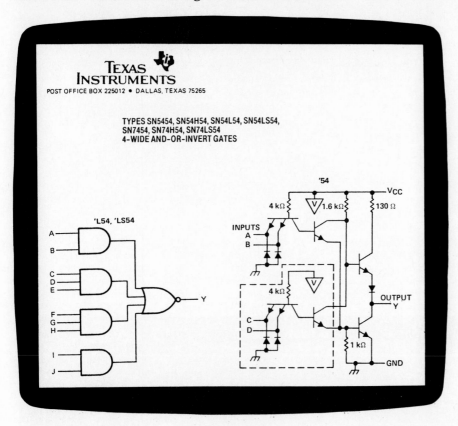

FIGURE P7-13

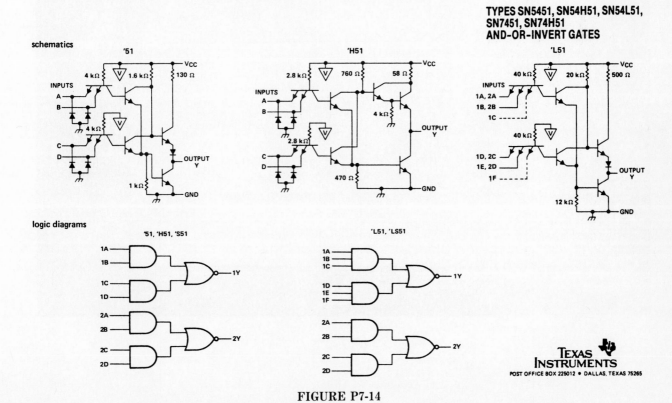

FIGURE P7-14

8

Integrated
Circuits

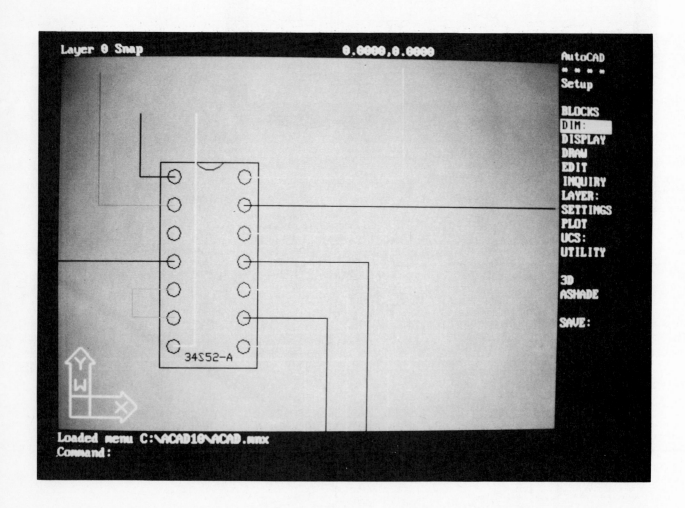

8-1 INTRODUCTION

Integrated circuits (ICs) are very small circuits etched into silicon wafers. The circuits are so small that 18 N-channel inverter circuits which contain two transistors each can fit under the period at the end of this sentence. Figure 8-1 shows an integrated circuit.

In this chapter we explain how integrated circuits are manufactured and what drafting procedures are required in the manufacturing process. Some IC design concepts are presented, but only in the depth needed to help understand how to create appropriate masks.

There are several terms which are used when speaking about ICs.

Bit   Binary digit, the smallest part of an IC. It has either high voltage or low voltage, expressed as either 1 or 0.

Byte   8 bits.

EPROM   Erasable programmable read-only memory.

Gate   The controlling element of certain transistors, or a basic logic circuit.

K   1024 bits (the value comes from the binary system). A 16K system has 16 × 1024, or 16,384 bits of information.

FIGURE 8-1 Integrated circuit. (Courtesy of Texas Instruments Inc.)

Logic   The order in which an IC is designed so as to produce certain outputs from given inputs.

LSI   Large-scale integration.

Mask   Plates used during manufacturing to define the required logic pattern.

Memory Chip   An IC that stores information using electrical charges.

Microprocessor   An IC that performs many different functions.

RAM   Random access memory; memory that can be changed by an operator.

ROM   Read-only memory; memory that can not be changed (e.g., a digital calendar).

Transistor   Functions as either an amplifier or a current switch in an IC.

VLSI   Very large scale integration.

Wafer   The thin disk on which an IC is manufactured.

8-2   HOW AN IC IS MANUFACTURED

An IC is manufactured on a thin piece of silicon called a wafer. The manufacturing of silicon wafers is called "growing" the purified silicon. The wafers are then cut into the desired shape and coated, first with an oxide, then with a coating of photoresist (see Figure 8-2).

FIGURE 8-2 How a silicon wafer is prepared for IC manufacture.

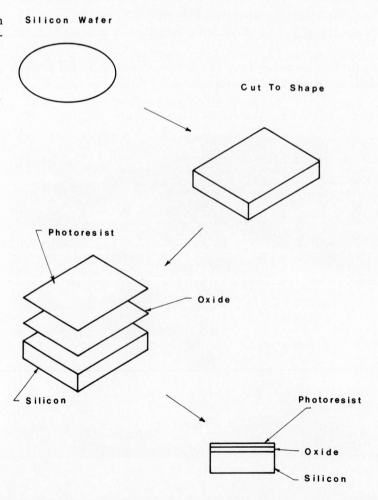

Photoresist material is light sensitive. If light is shown on photoresist material, the material becomes hard and will resist the acids and solvents used in etching. Conversely, the photoresist material *not* exposed to light (covered by a mask) will remain soft and can be removed by etching.

The treated silicon wafer is then exposed to light as shown in Figure 8-3. Areas exposed to light will harden. Areas blocked from the light by the mask will remain soft. The coated wafer is then etched, removing the soft photoresist.

The oxide layer is now processed by hot gases. The gas will not penetrate the hardened photoresist, but will enter the open areas created during etching, resulting in cavities in the oxide layer as shown.

The process is then repeated by first removing the hardened photoresist material and applying a new layer of silicon. A new layer of photoresist and oxide is attached, then etched as needed. The result will be a gradual buildup of cavities and filled areas which serve to form the IC chip.

FIGURE 8-3   How an IC is manufactured.

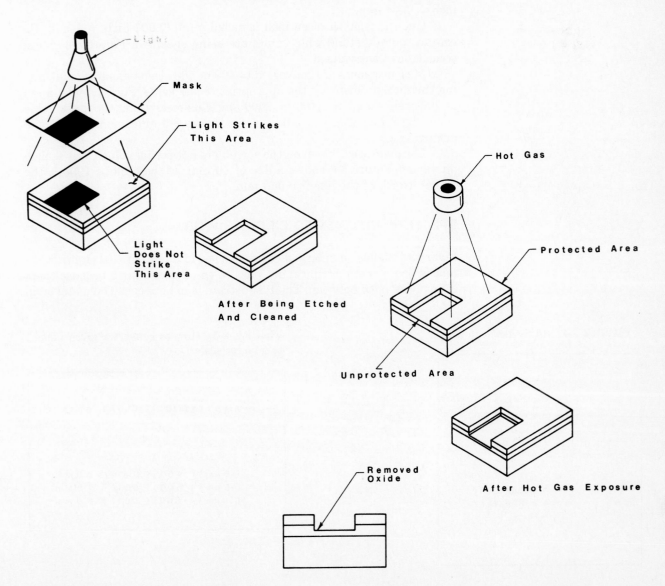

The conductor zones are interconnected by again coating the wafer with silicon and photoresist. The open areas that penetrate the silicon layers to the conductor zones are called *windows* and are created using a masking technique. Metal is then condensed into the windows to form the required connections between positive and negative zones.

The entire process is dependent on correctly prepared masks. Drafters are usually responsible for mask preparation.

8-3   CHOOSING A SCALE

Masks for IC manufacturing are done at scales from 200 to 1000:1, with 400 or 500:1 being most common. Very large scales are required because the final sizes are so small.

Measurements for IC masks are usually done using millimeters (mm). A millimeter is one thousandth of a meter. There are 10 millimeters per centimeter and 100 centimeters in a meter.

IC mask layouts use sizes measured in micrometers (μm). One thousandth of an inch equals twenty-five and four-tenths micrometers (0.001 inch = 25.4 μm).

One thousandth of an inch is called a mil (0.001 inch = 1 mil). Be careful to understand which units are being specified. Figure 8-4 shows some equivalent values.

Measurements are converted to 400 or 500 scale values by multiplying the measurement by the appropriate scale value. For example, 5 μm at 400 scale would be 5(400) = 2000 μm. This means that the mask value would be 2 mm (2000 × 0.001 = 2). Figure 8-5 gives some additional examples.

The distances shown on the screen are not actual values, but scaled-up values. Figure 8-6 shows a line of 40 mm. If the scale is 1000:1, the actual length of the line is 0.040 mm.

8-4   HOW INTEGRATED CIRCUITS WORK

Figure 8-7 shows an enlarged cutaway view of an integrated circuit. Current in the A metal strip cannot pass to the B metal strip because there is no direct path between. The P substrate is in between. The metal strip

FIGURE 8-4   Equivalent values.

$$1\ \mu m = 1 \times 10^{-6}\ m$$
$$1\ mm = 1 \times 10^{-3}\ m$$
$$25.4\ mm = 1.00\ in$$
$$0.001\ in = 2.54\ \mu m$$

$$\mu = micro$$
$$m = meter$$
$$mm = millimeter$$
$$in = inch$$

FIGURE 8-5   How to convert a given value to a larger scale.

$$(value)(scale)(.001) = mask$$

$$(5\ \mu m)(400)(.001) = 2.0\ mm$$
$$(15\ \mu m)(500)(.001) = 7.5\ mm$$
$$(8\ \mu m)(1000)(.001) = 8.0\ mm$$

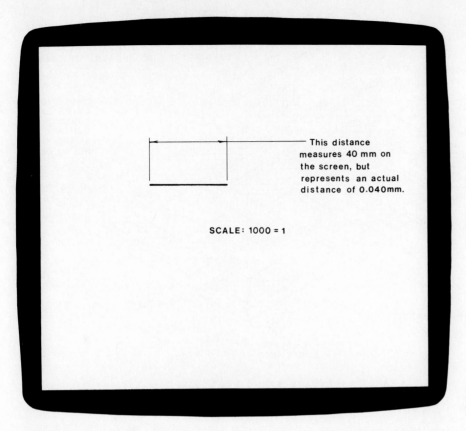

This distance measures 40 mm on the screen, but represents an actual distance of 0.040mm.

SCALE: 1000 = 1

FIGURE 8-6 How to draw a line at a scale of 1000 = 1.

FIGURE 8-7 How an IC MOS transistor works.

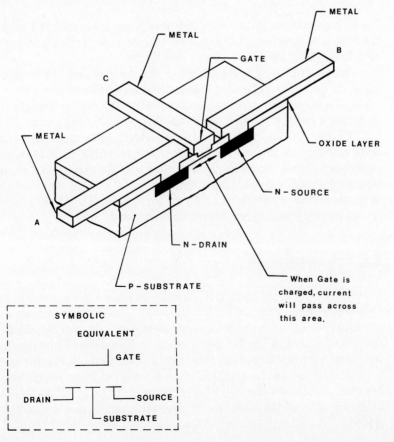

D: EX2

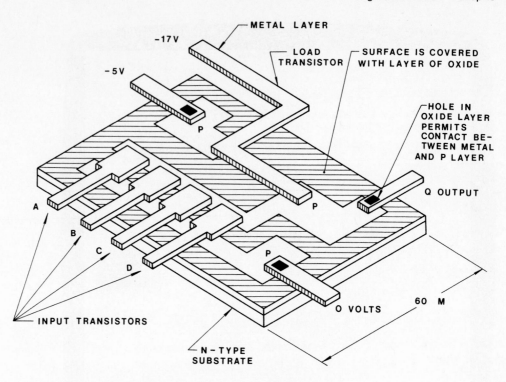

FIGURE 8-8   Four-input NAND gate structure.

C acts as a control gate. When current is applied to C, the area between the N-source and N-drain becomes electronically active and permits current to pass between the source and drain.

When current passes between the source and the drain, there is some current loss. The larger the distance between the source and the drain, the larger the current loss. The setup shown can therefore be made to act as a resistor as well as a switch.

Figure 8-8 shows a four-input NAND gate structure. This setup adds transistors in parallel to allow for multiple functions within the same very small structure. These very simple structures comprised of layers of silicon, oxide, N material, and metal can be repeated thousands of times within a single IC chip to form complex circuits that in turn create memory and other computer functions.

8-5 TRANSISTORS

Transistors are developed in IC circuits as shown in Figure 8-9. The figure shows the symbol for an NPN transistor. The three terminals c, b, and e are also labeled with their IC equivalents: N^+, P, and N^+. Step 1 shows the development of the N^+ and P regions. The dimensions for the regions will vary with speed requirements and voltage intensity for the circuit.

Step 2 shows the addition of an N^+ buried layer which helps reduce resistance between the N epilaxial layer and the P substrate. The N epilaxial layer is added after the N^+ buried layer. A new oxide layer is also added.

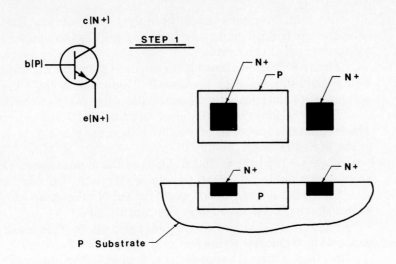

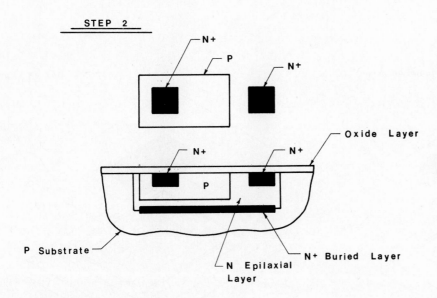

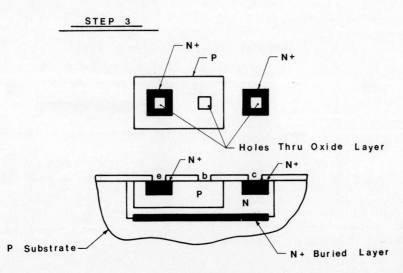

FIGURE 8-9 Development of an NPN transistor in an IC.

In step 3, holes are added through the oxide layer. These holes are equivalent to the c, b, and e terminals and are used to connect the transistor to the rest of the circuit.

Figure 8-10 shows how masks are used to create a transistor. Figure 8-10(a) shows an isolation area mask. This mask isolates the transistor and prevents electrical interference with other parts of the circuit.

Figure 8-10(b) shows the mask used to create the N$^+$ buried layer. The N epilaxial layer does not require a separate mask. It is added on top of the buried layer.

Figure 8-10(c) shows the addition of the P base layer. The isolation area is repeated in this mask to help reinforce the isolation area.

Figure 8-10(d) shows the mask for the N$^+$ areas, and Figure 8-10(e) shows the mask for the windows or contact hole.

Figure 8-10(f) shows the metalization mask. This mask is used to connect the transistor to the rest of the circuit. Metal, usually aluminum, is added in this step. The runners are broken in the mask, as they would continue into other areas of the circuit.

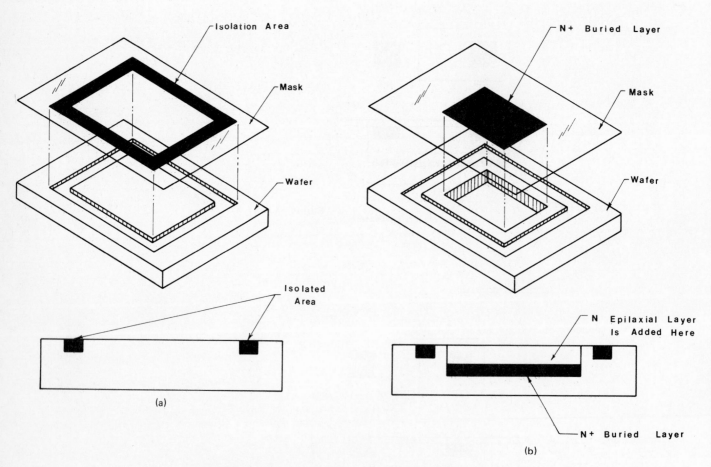

FIGURE 8-10   (a) Isolation area mask, (b) N$^+$ buried-layer mask; (c) P base layer mask; (d) N$^+$ layer mask; (e) contact hole mask; (f) metalization layer mask.

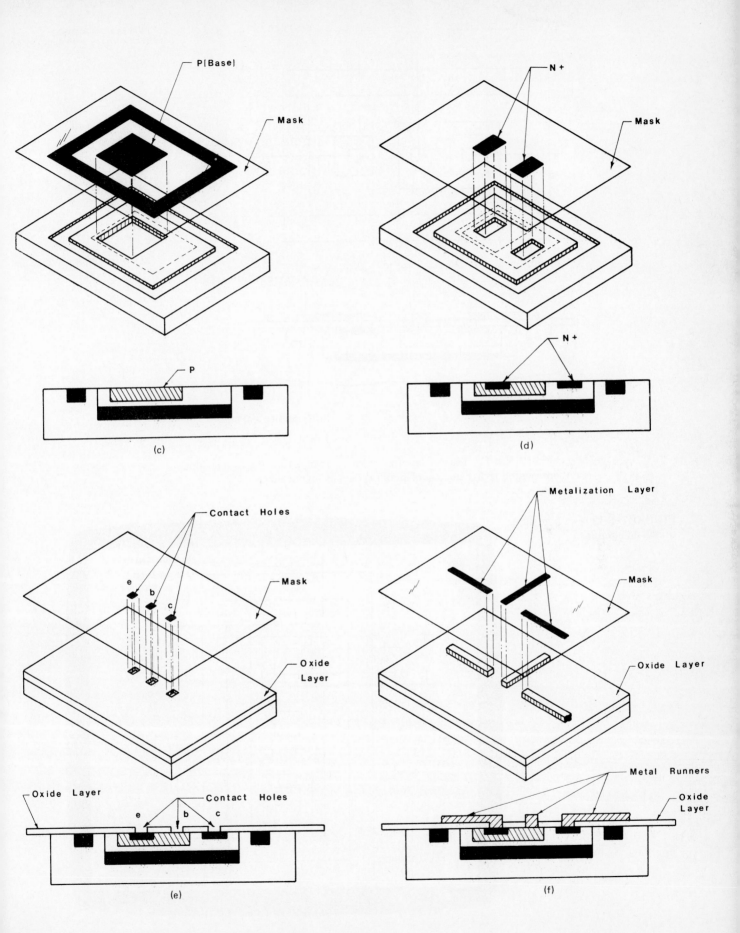

P (Base)

Mask

P

(c)

N+

Mask

N+

(d)

Contact Holes

Mask

Oxide Layer

e b c

Oxide Layer

Contact Holes

e b c

(e)

Metalization Layer

Mask

Oxide Layer

Metal Runners

Oxide Layer

(f)

161

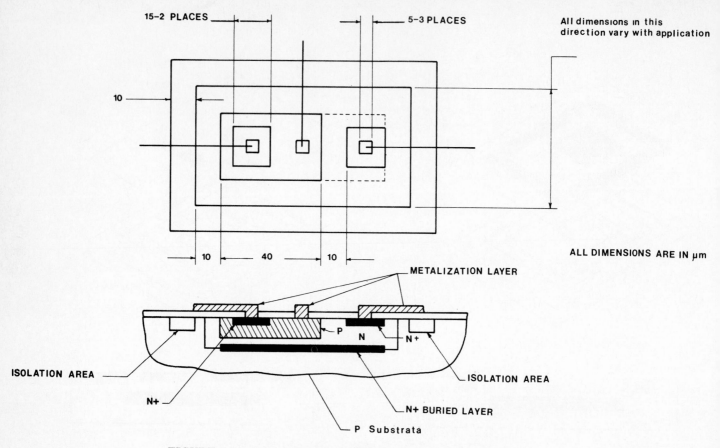

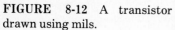

FIGURE 8-11 Master layout for an IC transistor. The dimensions presented represent one of many different layout sizes.

FIGURE 8-12 A transistor drawn using mils.

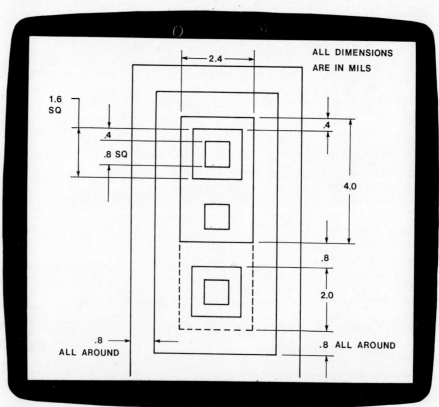

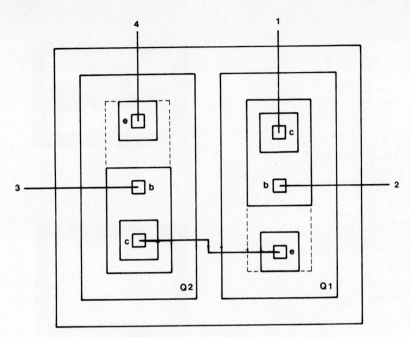
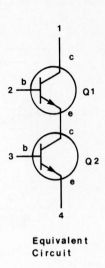

Equivalent
Circuit

FIGURE 8-13   Two NPN transistors.

Masks are created from a master layout as shown in Figure 8-11. The master layout is used to space the various layers. The hidden lines represent the buried layer. The dimensions given are only representative of a typical transistor layout. Dimensions will vary with application. Figure 8-12 shows a transistor master layout dimensioned in mils.

Transistors may be positioned next to each other as shown in Figure 8-13. Note in the equivalent circuit that the e of Q1 is connected to the c of Q2. This connection is shown as a step in the master layout. This is because most layouts are done on computer graphics systems which can draw only horizontal and vertical lines. Note that the two transistors are separated by an isolation area.

8-6   RESISTORS

Resistance in integrated circuits can be created using the base material, the emitter material, the collector material, or by a process called pinching. This section deals with base material resistance, which is the most widely used process. Figure 8-14 shows a base diffused resistor.

The masks needed to create the resistor are also shown. Note that an N$^+$ buried layer is also included and that the resistor is isolated. Further, the isolation is repeated in the buried layer mask to help to reinforce it.

The amount of resistance depends on the resistance of the material used, the material's cross-sectional area, and the material's length. The resistance of the material is called *sheet resistance* and is measured in *ohms per square* ($\Omega/\square$). The total value of a resistor is calculated using the formula

$$R = R_s \left(\frac{l}{w} \right) + \text{corners} + \text{contacts}$$

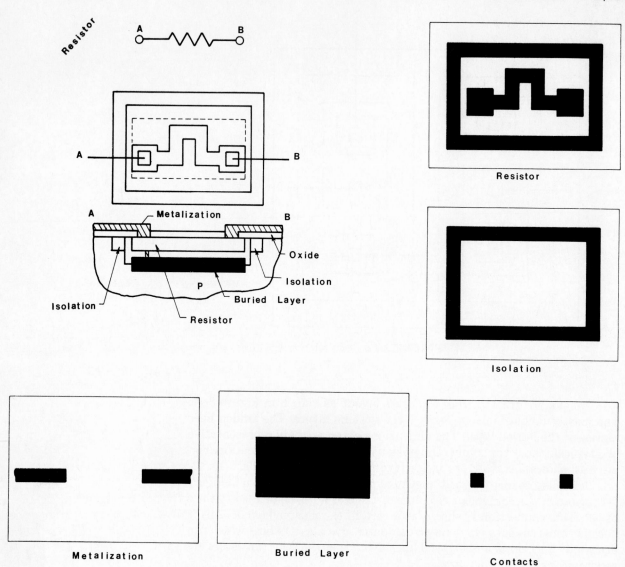

FIGURE 8-14   Resistor in an IC together with the appropriate masks.

where R_s = sheet resistance
l = length of sheet
w = width of sheet

Figure 8-15 shows these terms. For simplicity, corners and contacts generally are valued at one-half a sheet square. So, for example, if the sheet resistance is 200 Ω/□, the corner and the contact area would each be worth 100 Ω/□.

Figure 8-16 shows a resistance between points A and B. The value of the resistance is calculated as follows:

$$R = R_s \left(\frac{l}{w} \right) + \text{corners} + \text{contacts}$$

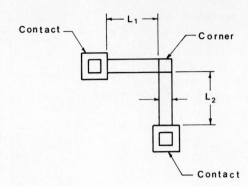

FIGURE 8-15   Resistor terms.

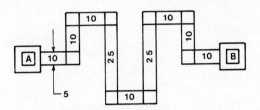

FIGURE 8-16   Resistance between points A and B.

Given $R_s = 200 \ \Omega/\square$, we have

$$R = (200) \frac{10 + 10 + 10 + 25 + 10 + 25 + 10 + 10 + 10}{5}$$

$$+ \ 8 \text{ corners} + 2 \text{ contacts}$$

$$= 4800 + 800 + 200$$

$$= 5800 \ \Omega$$

Figure 8-17 shows a circuit that has a transistor and two resistors. The master layout and masks have also been included. The transistor is isolated, but both resistors are located within the same region. The metal runners cross between the isolation area connecting the various terminal windows.

8-7   DIODES AND CAPACITORS

Figure 8-18 shows an IC diode both as a symbol and as it would appear on an IC chip. The various layers are created using the masking techniques shown for transistors and resistors. Diodes with common anodes or cathodes may be combined to save space. Figure 8-18 shows an example of a master layout for two diodes with a common cathode.

Figure 8-19 shows an approximate representation of a master layout for a capacitor. There is a continuous relationship between capacitors and diodes in IC chips called a parasitic relationship. The actual design and subsequent layout of capacitors is beyond the scope of this book.

FIGURE 8-17 IC that has two transistors and a resistor, together with the appropriate masks.

SCHEMATIC DIAGRAM

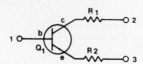

MASTER LAYOUT

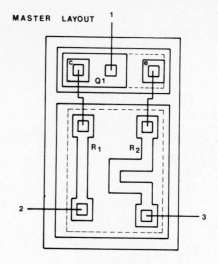

MASKS

Isolation

Buried Layer

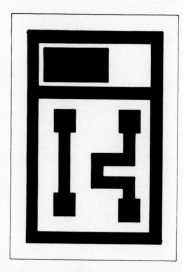

BASE

EMITTER

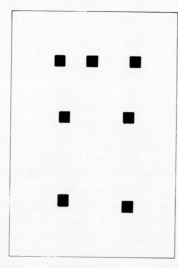

CONTACT

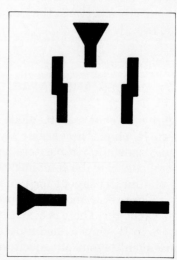

METALIZATION

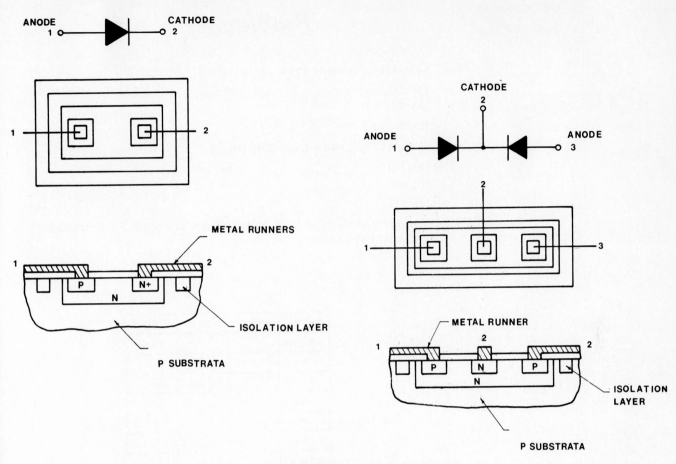

FIGURE 8-18   Diodes in an IC.

FIGURE   8-19 Approximate capacitor in an IC.

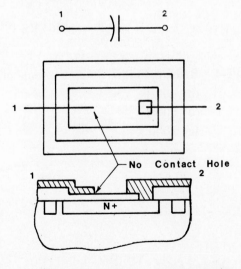

THIS IS ONLY AN APPROXIMATE REPRESENTATION

Problems

P8-1 Convert each value from μm to mils.

    (a) 10   **(d)** 15

    (b) 50   **(e)** 20

    (c) 5

P8-2 Convert each value from mils to μm.

    (a) 1.0   **(d)** 1.5

    (b) 2.0   **(e)** 2.5

    (c) 1.7

P8-3 Redraw and complete the table shown in Figure P8-3. Include units for each value.

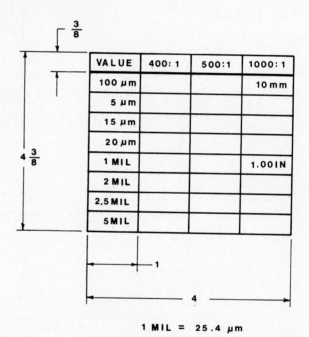

VALUE	400:1	500:1	1000:1
100 μm			10 mm
5 μm			
15 μm			
20 μm			
1 MIL			1.00 IN
2 MIL			
2.5 MIL			
5 MIL			

1 MIL = 25.4 μm

FIGURE P8-3

P8-4 Redraw Figure P8-4 at a scale of:

    (a) 400:1

    (b) 500:1

    (c) 1000:1

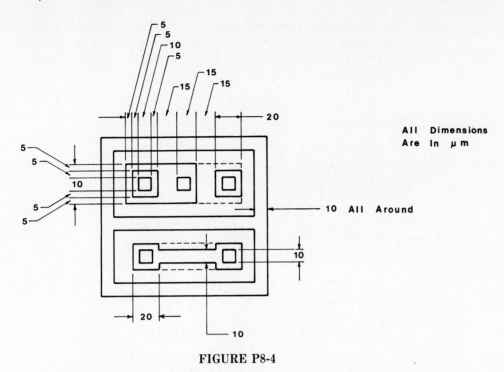

FIGURE P8-4

P8-5   Calculate the resistance for the shape shown in Figure P8-5. Assume that the material has a sheet resistance of 200 $\Omega/\square$.

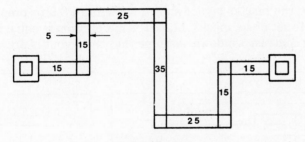

FIGURE P8-5

P8-6   Calculate the resistance for the shape shown in Figure P8-6. Assume that the material has a sheet resistance of 200 $\Omega/\square$.

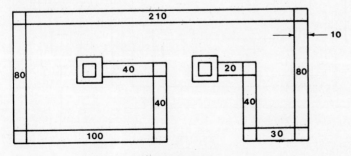

FIGURE P8-6

P8-7   Design a resistance pattern that generates a resistance of 20 kΩ ± 10%. Assume that the two end points, A and B, are located 100 μm from each other.

P8-8   Use the master layout shown in Figure P8-8 to prepare a complete set of six masks. Make the metal runners equal in width to the contact windows. Assume that the layout is drawn to scale.

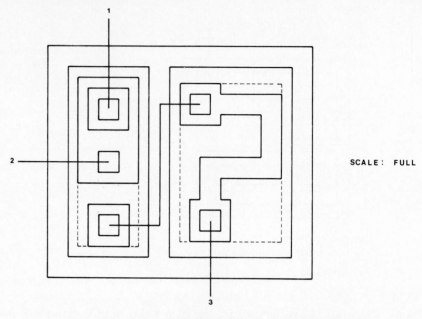

SCALE: FULL

FIGURE P8-8

P8-9   Use the master layout shown in Figure P8-9 to prepare a complete set of six masks. Make the metal runners equal in width to the contrast windows. Assume that the layout is drawn to scale.

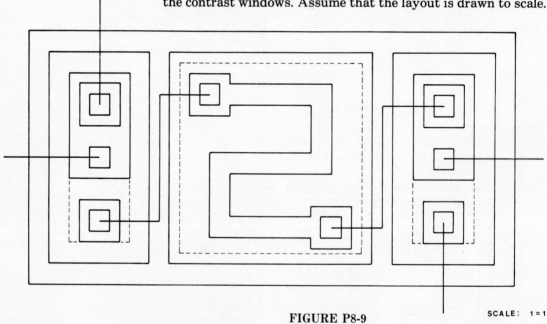

FIGURE P8-9   SCALE: 1 = 1

P8-10   Make a complete set of five masks from the master layout of the resistor shown in Figure P8-10. Make the metal runners equal in width to the contact windows. Assume that the layout is drawn to scale.

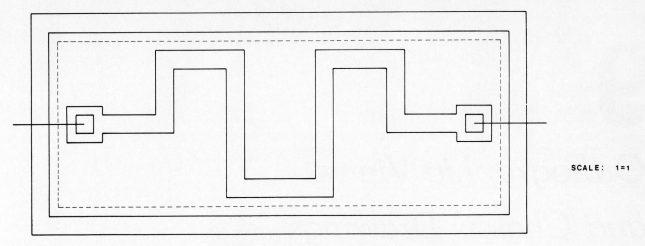

FIGURE P8-10

P8-11 Prepare a master layout and appropriate masks (a set of six) for the schematic diagram shown in Figure P8-11. Use the general dimensions shown in Figure P8-10 to size the transistors. Assume that the resistance material is 10 μm wide and has a sheet value of 200 $\Omega/\square$. Assume that the end area is 15 μm square.

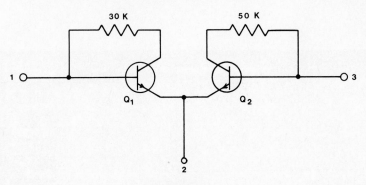

FIGURE P8-11

P8-12 Repeat Problem P8-11 for the schematic shown in Figure P8-12.

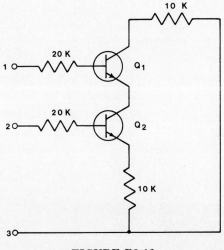

FIGURE P8-12

9

Orthographic Views and Chassis Drawings

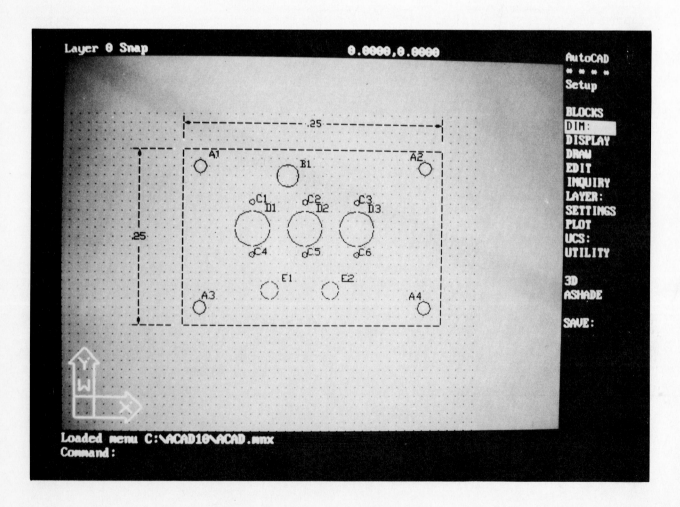

9-1 INTRODUCTION

This chapter introduces the fundamentals of orthographic views and shows how they are applied to chassis drawings. Also included is an explanation of sheet metal bending and how to dimension chassis.

9-2 ORTHOGRAPHIC VIEWS

Orthographic views are two-dimensional drawings of an object taken at fixed angles relative to the object. Figure 9-1 shows an object and one possible orthographic view of the object.

The orthographic view shown in Figure 9-1 shows the height and width of the object, but not the depth. There is no information concerning the depth of the object on the given orthographic view. Additional orthographic views, taken at other angles around the object, are needed to define the shape completely, including depth.

The two-dimensional limitations of orthographic views mean that line 1-2, which represents one of the edges of the object shown in Figure 9-1, appears as a point (corner) in the orthographic view. This point is an *end view* on line 1-2. Plane *A* in Figure 9-1 appears as a line in the orthographic view. The line is an end view of the plane.

The concept of end views may be visualized by holding a sheet of paper and rotating it until the plane of the paper is directly in line with

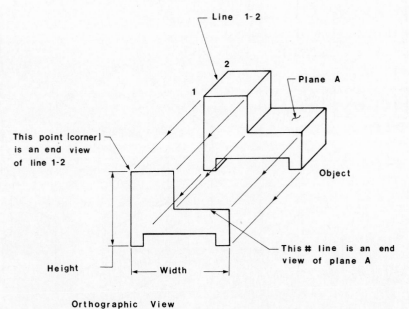

FIGURE 9-1 Orthographic view of an object.

173

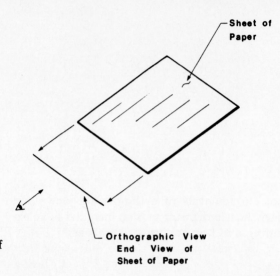

FIGURE 9-2 The end view of
a plane is a straight line.

your line of sight. If the sheet of paper is aligned directly behind the edge,
the sheet will appear as a single edge line. The paper edge is the same as
an orthographic view taken in line with the plane of the paper (Figure
9-2).

Orthographic views may be taken at any angle to an object. Assume
that the object was within a box as shown in Figure 9-3, and that an
orthographic view was projected from the object to each face of the box.
The box faces could then be unfolded to produce the six views shown.

FIGURE 9-3   Six orthographic views of an object.

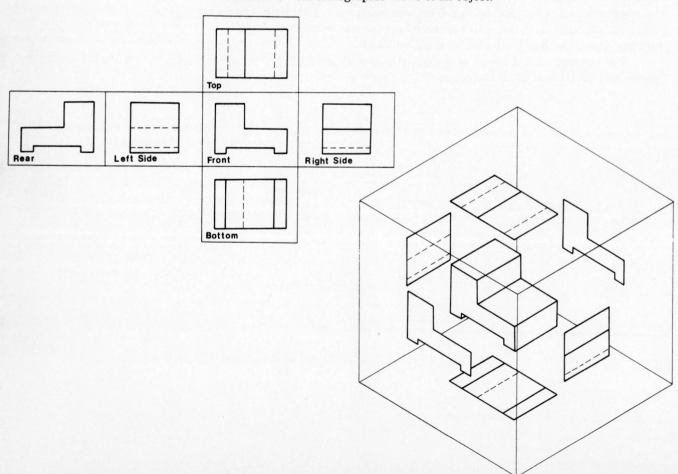

Each of the six views was taken at 90° to every other. In this chapter we explain how orthographic views are generated, how they relate to one another, and how they are interpreted.

9-3   THREE VIEWS OF AN OBJECT

Standard drafting convention calls for three orthographic views to define an object: a front, top, and right-side view. Some objects may require more than three views, and some fewer, for complete shape definition, but the primary approach to defining an object using orthographic views should always be three views.

Figure 9-4 shows an object and the front, top, and right-side views of the object. The location of the three views, relative to each other, is critical. The top view must be located directly over the front view, and the right-side view directly to the right of the front view. Each view of a feature must be directly aligned with the same feature in the other orthographic views. For example, the top view of surface A in Figure 9-4 is located directly over the front view of surface A. The right-side view of surface A is located directly to the right of the front view. Projection lines have been added in Figure 9-4 to help accent the alignment. Figures 9-5 and 9-6 are further examples of objects and three orthographic views of the objects.

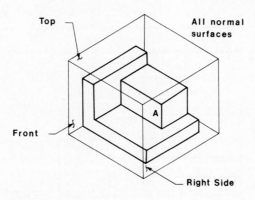

FIGURE 9-4 Standard three orthographic views of an object: front, top, and side.

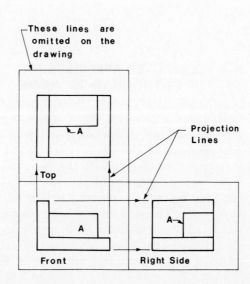

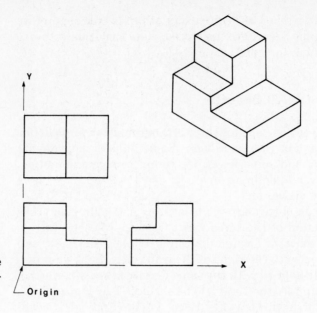

FIGURE 9-5 Object and three orthographic views of the object.

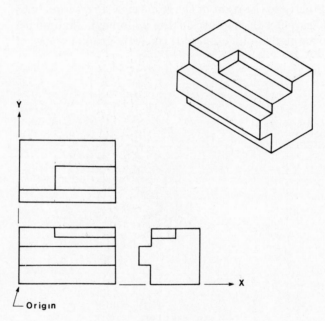

FIGURE 9-6 Object and three orthographic views of the object.

9-4 COMPUTER APPLICATIONS OF ORTHOGRAPHIC VIEWS

Orthographic views are drawn on a computer graphic using one of several different methods, depending on the system's capability. One system is first to position the front view relative to the XY axis. The other two views are then aligned with the front view along the XY axis system. The front view defines the height and length of the object (Figure 9-7). The top view defines the object's length but is coupled with the depth measurements. The side view shares the height values with the front view, but in addition, includes the depth values.

The distance between views is usually equal. This helps give the drawing a more organized appearance. The distances between views need

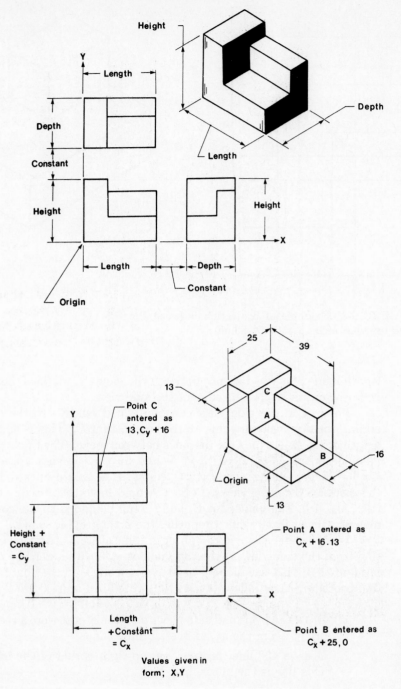

FIGURE 9-7   How to locate orthographic views relative to an *XY* axis system.

not be the same, and if extra space is needed for dimensions, the distances may be increased. A good rule of thumb is about 1.00 in. or 25 mm between views.

A consistent distance between views is helpful when using a computer graphic system. The distance can be treated as a constant and can be factored into the view locations. For example, in Figure 9-7 the overall values for length, height, and depth are known. The distance between views is also known. This means that coordinate points used to define the top and side views can be taken directly from given dimensions by adding the value of the distance between the views. For example, C_x equals the

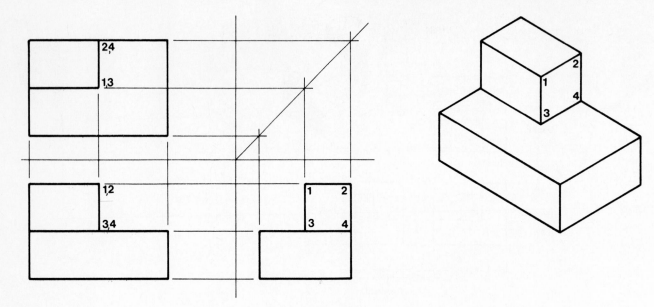

FIGURE 9-8 Example of projection between the side and top view using a 45° miter line.

length value plus the distance between the views. C_x defines the distance from the origin to the left edge of the side view.

Point A in the side view can be defined relative to the origin by adding the dimensional value 16 to the constant C_x. This is equal to the length of the object plus the distance between views. The final coordinate value becomes $C_x + 16, 13$ (x, y). Values for the top view are found in a similar manner, using a constant C_y equal to the height of the object plus the distance between views. Point C would have the coordinate values (10, $C_y + 16$). The values for C_x and C_y can be added to the computer's memory and utilized while preparing the orthographic views.

It is possible on some computer graphic systems to project information from the side view to the top view using a 45° miter line as shown in Figure 9-8. This technique is used extensively in on-the-board drafting. The procedure is as follows and is illustrated in Figure 9-9.

1. Draw the front and side views as if the object were a rectangular prism using the given overall dimensions.

2. Draw a 45° line through the top right corner of the front view. This line is the miter line.

3. Complete the front and side views and draw vertical lines through all the corners in the side view so that they intersect the miter line. Define the intersection points. (*Note:* This method is dependent on the accurate location of these intersection points. Some systems can define a point by identifying two intersecting elements; others need specific coordinate inputs. If a cursor or light pen is used to locate the points by eye, the locations may not be accurate. It is suggested that if the points are to be located by eye, the drawing scale be increased to its maximum size before trying to define the points, so as to minimize location errors.)

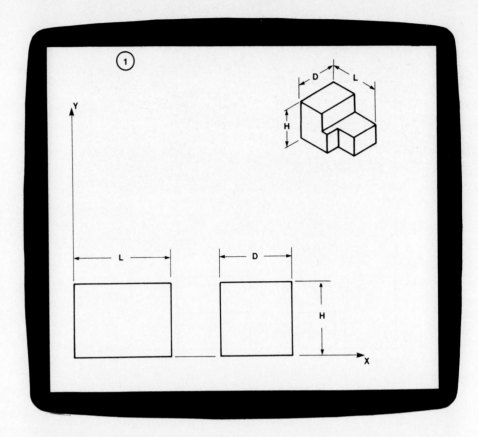

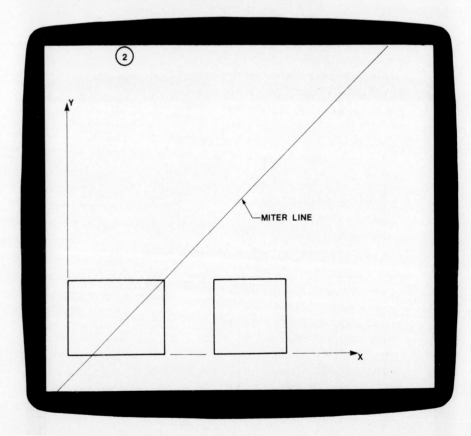

FIGURE 9-9   How to use a 45°
miter to project between views.

FIGURE 9-9 (*cont'd*)

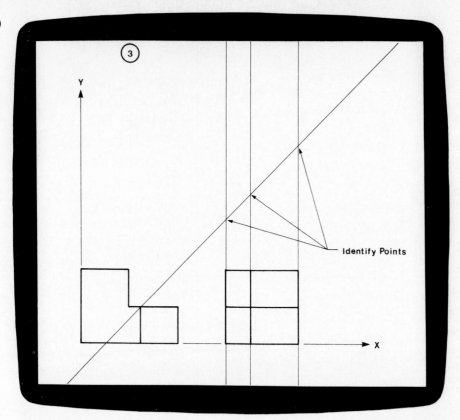

4. Draw horizontal lines through the intersection points of step 3.

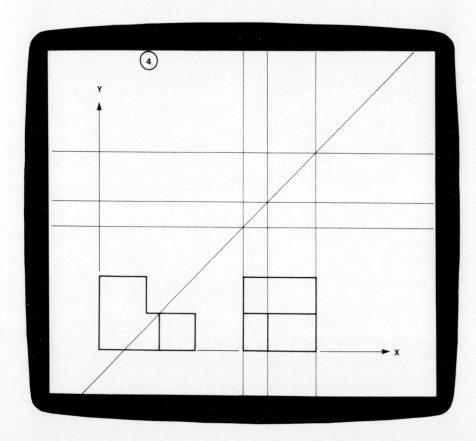

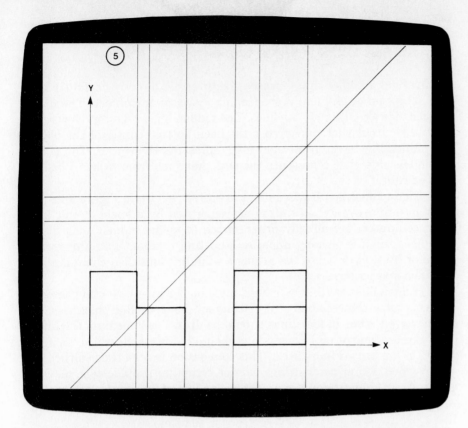

FIGURE 9-9 (*cont'd*)

5. Draw vertical lines through all corners in the front view. The intersection of these lines with the horizontal lines of step 4 will define the lines in the top view.

6. Erase all excess lines and change all needed lines to their correct thickness and font.

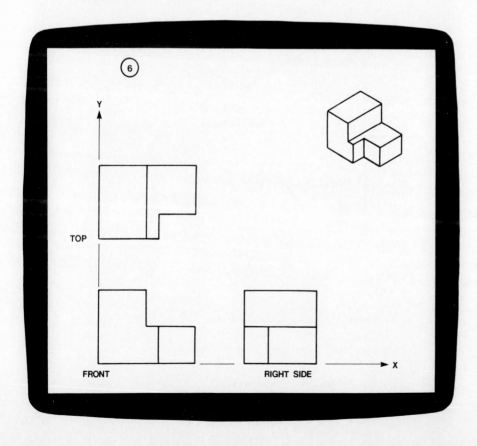

9-5 HIDDEN LINES

Hidden lines are lines that represent features not directly visible. Figure 9-10 shows an object and orthographic views of the object. The orthographic views contain hidden lines. The hidden line in the top view represents the cutout that runs across the back bottom corner of the object. The hidden line in the front view also represents the cutout. The hidden line in the side view represents the slot that runs from front to back in the top center of the object.

Hidden lines are drawn using the pattern shown in Figure 9-11. The lines are approximately 4 units long separated by a space 1 unit long. Most computers are already programmed to create hidden lines to this specification. It is usually done by inputting a hidden-line font from a menu or by first drawing the line as a solid line and then changing it to a hidden-line pattern.

Hidden lines are drawn using a medium thickness. Not all computer systems have the capability to distinguish between line thickness, but whenever possible, hidden lines should be drawn at a medium thickness.

Corners and other intersections should be shown clearly, even if hidden lines are used (Figure 9-12). This convention is sometimes difficult to follow when preparing drawings using a computer graphic system. Computers draw hidden lines at a predetermined fixed line length and spacing

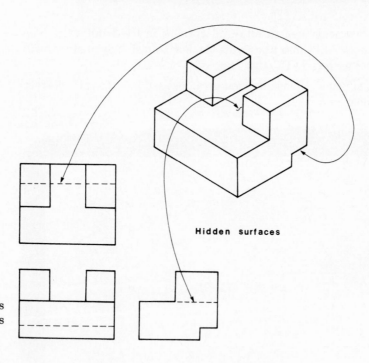

Hidden surfaces

FIGURE 9-10 Hidden lines are used to represent features that are not directly visible.

FIGURE 9-11 Pattern for hidden lines.

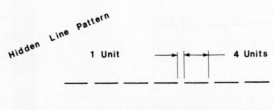

Hidden Line Pattern

1 Unit   4 Units

Medium Thickness

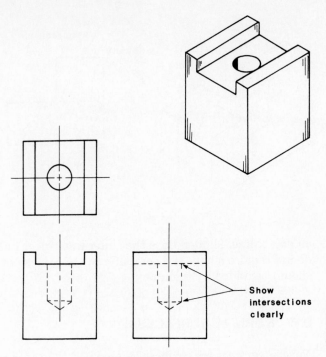

FIGURE 9-12 Corners and intersections must be shown clearly.

distance. This makes it difficult to assure correct intersections. Drafters often vary the pattern slightly to assure clear corners and intersections. If possible, always show all hidden intersections clearly.

Parallel hidden lines that are very close together should be staggered to prevent confusion (Figure 9-13). This convention is also difficult to achieve on a computer system that uses predetermined line and spacing distances. Staggering can be created by starting one of the lines at some point beyond the actual starting point and then trimming any excess. In Figure 9-13, the second line was intentionally drawn beyond the needed limits to create the staggering effect. The excess can be trimmed.

Object lines should never run directly into a hidden line (Figure 9-14). A small gap is required to help distinguish between the two types of lines. Gaps are needed *only* when a hidden line is directly aligned with

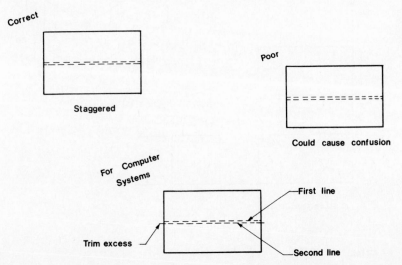

FIGURE 9-13 Stagger the spacing of close parallel hidden lines to ensure clarity.

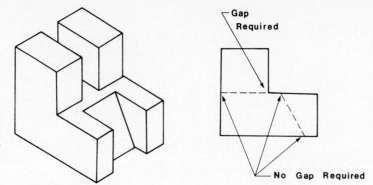

FIGURE 9-14 Object lines should never run directly into hidden lines. The two types of lines should be separated by a small gap.

an object line. Hidden lines that intersect object lines at any other angle do not require a gap. If a computer system does not allow for gaps, they should be added by trimming.

9-6 NORMAL SURFACES

Normal surfaces are surfaces located 90° from each other. A normal angle is a 90° angle.

All lines in orthographic views of objects that contain only normal surfaces are *true length* and all surfaces are *true shape*. This means that the front view of the object shown in Figure 9-15 is drawn as a rectangle 50 × 100 mm as specified by the given dimensions. Similarly, the side view is drawn as a square, 50 × 50 mm.

All lines in Figure 9-15 are either parallel or perpendicular to the X and Y axes. Figure 9-16 shows another object that contains only normal surfaces. Note again that all lines are parallel to either the X or Y axis. All lines are true length and all surfaces are true shape.

FIGURE 9-15 Normal surfaces are located 90° from each other. All surfaces are true length and shape.

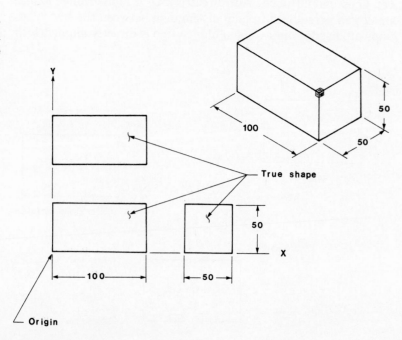

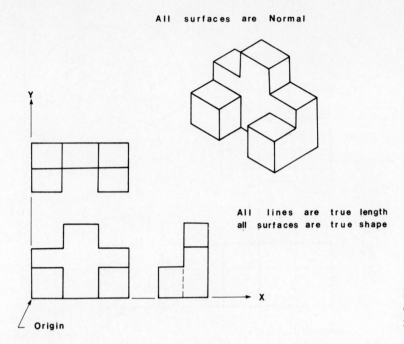

FIGURE 9-16 Example of an object that contains only normal surfaces.

9-7 SLANTED SURFACES

Slanted surfaces are surfaces drawn at an angle to the XY axis. Figure 9-17 shows an object that contains a slanted surface. The front view shows a profile of the slanted surface (an end view of the surface). The surface appears as a rectangle in both the top and side views, but these rectangles are *not* true shapes. Both are foreshortened from their actual size. The size on the slanted surface in the top and side view is determined by projection.

Projection lines are parallel to either the X or Y axis as shown. This ensures that the views will be aligned as required by standard convention.

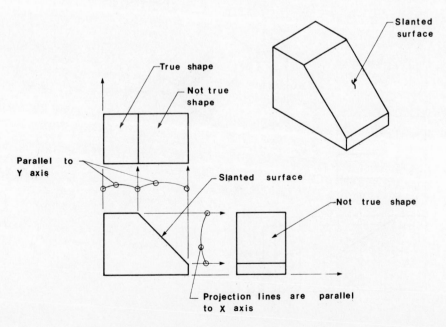

FIGURE 9-17 Object that contains a slanted surface.

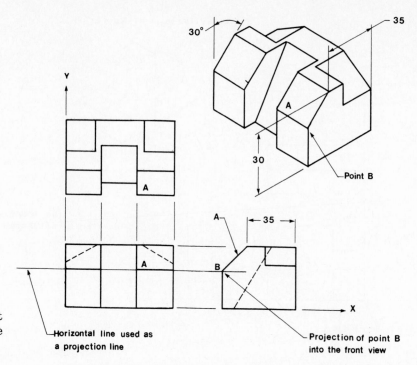

FIGURE 9-18 How to project point *B* from the side to the front view.

Projection lines are useful when preparing orthographic views of more complex objects. For example, in Figure 9-18 the surface labeled *A* is a slanted surface. The dimensions define the slant as 30° to the vertical and the overall height is 30 mm. This information is sufficient to draw the side view but not the front view. The vertical distance of point *B* from the top surface is needed. This distance could be calculated but is more easily determined by projection. Because the front and side views are aligned, the vertical distance can be projected using a horizontal line of unlimited length through point *B* in the side view, across the front view.

FIGURE 9-19 How to project information between views.

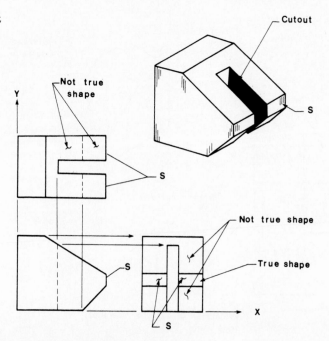

Projection lines are primarily construction lines, but serve, in part, to define object lines. The excess may be trimmed, leaving only those lines required for a correct front view.

Figure 9-19 shows another example of an object with slanted surfaces. The front view was chosen because it gives a true picture of the various angles involved in the object's shape. The top and side views of the slanted surfaces are not true shapes. The intersection of the cutout with the slanted surfaces was found by projection. Note that surface *S* is shown in true shape in the side view because its front view is parallel to the *Y* axis (it is a surface normal to the *XY* axis).

9-8 ROUNDED SURFACES

Rounded surfaces are surfaces whose end view is an arc of constant radius. Figure 9-20 shows an object that contains a rounded surface. Note how the top and side views show the rounded surface as a rectangle. There is no indication in these views that the surface is round.

Figure 9-21 shows a convex rounded surface. The rounded surface becomes tangent (blends smoothly into) to both the vertical and horizontal surfaces of the object. There is no edge line between the two surfaces. This means that the top and side views will appear rectangular with no indication of the shape of the rounded surface and with no edge line to show where the two surfaces become tangent.

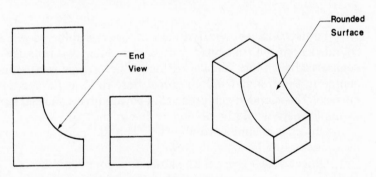

FIGURE 9-20 Object with a rounded surface.

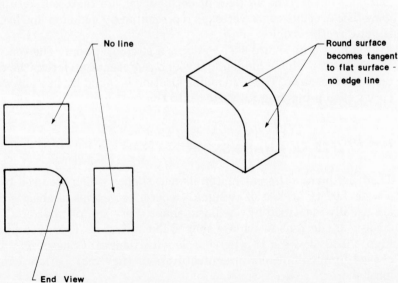

FIGURE 9-21 Object with a convex rounded surface.

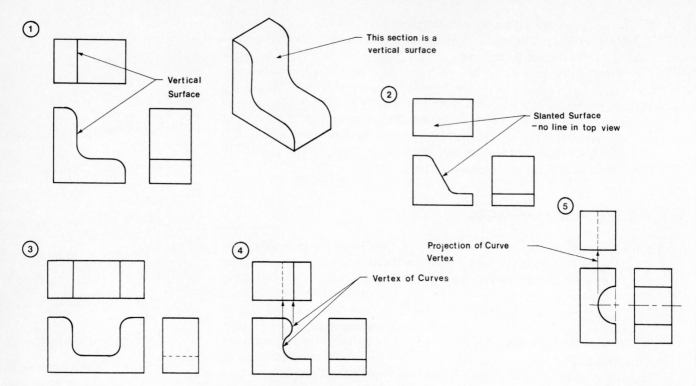

FIGURE 9-22   Five sets of orthographic views that contain rounded surfaces.

Figure 9-22 shows five sets of orthographic views that contain rounded surfaces. In object 1 of Figure 9-22, two rounded surfaces are separated by a short vertical surface. This vertical section must be represented in the top view by an object line. In object 2 there is no vertical surface. The rounded surfaces are separated by a slanted surface. In this example there would be no object line in the top view.

Object 3 is similar to object 1. The flat horizontal surface appears as a hidden line in the side view.

Object 4 in Figure 9-22 shows two rounded surfaces, one concave and one convex. The vertices of both curves are represented in the top view. The concave curve vertex is represented by a hidden line because it is not directly visible.

Object 5 in Figure 9-22 contains a circular cutout. The top view of the object contains a hidden line which represents a vertex (the deepest point) of the curve. The hidden-line location is found by first drawing the curve, then projecting its vertex into the top view.

9-9   TYPES OF CHASSIS

There are many different materials and shapes which are used to create a wide variety of chassis designs. General categories of chassis designs are usually identified by the basic shape. For example: box, U, T, and I chassis are designs which are shaped like a box, U, T, or I, respectively. Figure 9-23 shows a U-type chassis with flanges. Chassis may be purchased from commercial manufacturers or they may be fabricated in a local shop.

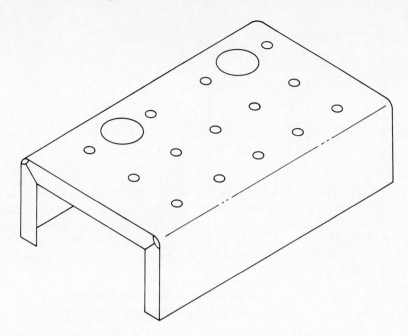

FIGURE 9-23 U-type chassis with flanges.

9-10 FLAT PATTERN DESIGN

Chassis are fabricated by first cutting a flat pattern from the desired material and then bending the flat pattern into the final chassis shape. As metal is bent, it stretches, which means that the length of the flat pattern must be less than the finished overall length of the chassis.

The correct size of a flat pattern is determined by considering the overall size requirements and the amount of material for *bend allowance*. Bend allowance is the amount of material needed to form bends and includes a consideration of how the material will stretch during the bending. Bend allowances and the lengths of straight sections are added together to determine the diminished length of the flat pattern.

Before proceeding with a discussion of how to determine flat pattern lengths, we must first understand three terms: *inside bend radius, outside bend radius,* and *neutral axis.* Figure 9-24 defines the terms. The neutral axis is a theoretical line exactly halfway between the inside bend radius

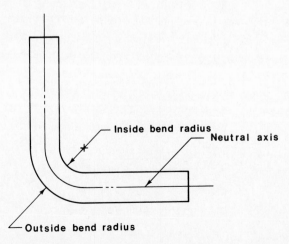

FIGURE 9-24 Illustration of inside bend radius, neutral axis, and outside bend radius.

Inside bend radius

Neutral axis

Outside bend radius

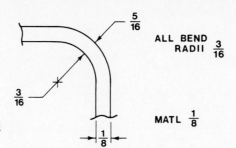

FIGURE 9-25 A sample bend radius.

and the outside bend radius. Material between the neutral axis and the inside bend radius is under compression during bending and material between the neutral axis and the outside bend radius is under tension (being stretched).

The difference between the inside and outside bend radii is equal to the material thickness:

$$\text{inside bend radius} + \text{material thickness} = \text{outside bend radius} \qquad \text{(F-1)}$$

Drawings usually define only inside bend radius and material thickness, so if the value of the outside bend radius is desired, it must be calculated using Formula (F-1). In Figure 9-25 the inside bend radius is $\frac{3}{16}$ inch and the material thickness is $\frac{1}{8}$ inch. Applying these data to Formula (F-1), we get

$$\frac{3}{16} + \frac{1}{8} = \text{outside bend radius}$$
$$\frac{3}{16} + \frac{2}{16} = \text{outside bend radius}$$
$$\frac{5}{16} = \text{outside bend radius}$$

The length of a flat pattern is determined by adding the length of the straight sections of the final chassis shape to the lengths of the curved sections, which have been modified to account for bend allowance. If we apply this concept to the contour shown in Figure 9-26, we see that

$$\text{flat pattern length} = d_1 + d_2 + d_3$$

The distance d_1 is found by taking the overall length and subtracting the inside bend radius and material thickness:

$$d_1 = 1.38 - 0.125 - 0.125$$
$$= 1.13$$

FIGURE 9-26 A sample bend radius.

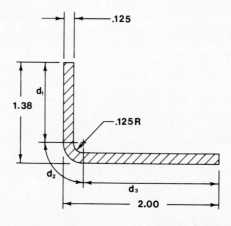

Similarly,

$$d_3 = 2.00 - 0.125 - 0.125$$
$$= 1.75$$

The distance d_2 must be calculated considering the bend allowance. This is done using the formula*

$$B = \frac{A}{360} 2\pi(\text{IR} + Kt) \qquad\qquad (\text{F-2})$$

where B = bend allowance
    A = bend angle
    IR = inside bend radius
    K = constant
    t = material thickness

For the problem presented in Figure 9-26,

$$A = 90°$$
$$\text{IR} = 0.125$$
$$K = 0.33$$
$$t = 0.125$$

The value of K is equal to 0.33 when the inside bend radius is less than $2t$ (two times the material thickness) and is equal to 0.50 when the inside bend radius is greater than $2t$. In our problem,

$$2t = 2(0.125) = 0.25$$

which means that

$$K = 0.33$$

as the inside bend radius is 0.125, which is less than 0.25. Subtracting the values in Formula (F-2), we have

$$B = \frac{90}{360} (2\pi)[0.125 + (0.33)(0.125)]$$
$$= 1.57 (0.125 + 0.041)$$
$$= 0.26$$

Therefore,

$$d_2 = 0.26$$

We can now calculate the flat pattern length by adding d_1, d_2, and d_3:

$$\text{flat pattern length} = d_1 + d_2 + d_3$$
$$= 1.13 + 0.26 + 1.75$$
$$= 3.14$$

Figure 9-27 is another example of how to calculate the flat pattern length for a given chassis shape.

*From ASTM *Die Design Handbook*.

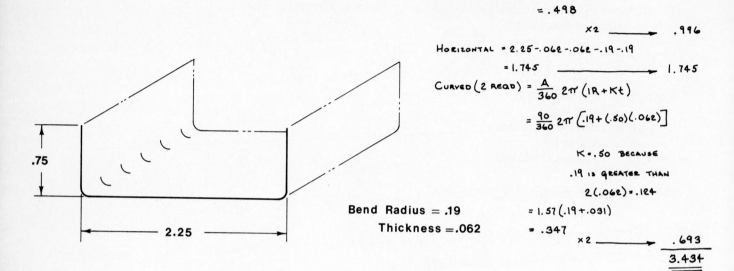

CALCULATION SHEET

STRAIGHT SECTIONS

VERTICAL (2 REQD) = .75 - .062 - .19

= .498

×2 ⟶ .996

HORIZONTAL = 2.25 - .062 - .062 - .19 - .19

= 1.745 ⟶ 1.745

CURVED (2 REQD) = $\frac{A}{360} 2\pi (IR + Kt)$

= $\frac{90}{360} 2\pi \left[.19 + (.50)(.062)\right]$

K = .50 BECAUSE

.19 IS GREATER THAN

2(.062) = .124

Bend Radius = .19   = 1.57(.19 + .031)

Thickness = .062   = .347

×2 ⟶ .693

3.434

.75

2.25

FIGURE 9-27   Example of a flat-pattern length calculation.

9-11   ORTHOGRAPHIC VIEWS OF CHASSIS

Orthographic views of chassis follow the fundamentals for orthographic views presented earlier in the chapter. Chassis are made from thin materials such as sheet metal. Figure 9-28 shows the orthographic views of a thin sheet metal part. Note the location and relative positions of the three views.

Holes in thin objects are drawn using hidden lines as shown. If the material is so thin that there is not enough space to draw hidden lines, just a centerline may be used to indicate the hole location, as shown in Figure 9-29. Figure 9-30 is a further example of orthographic views of an object made from thin material.

9-12   HOW TO DIMENSION CHASSIS DRAWINGS

Chassis designs are characterized by many small holes in a small area. To help produce neat, clear, and easy-to-read chassis drawings, two different dimensioning systems are commonly used: the baseline system and the coordinate system.

The *baseline system* refers all dimensions to common baselines. These lines are sometimes called datum lines or reference lines. The baseline system is illustrated in Figure 9-31. The baseline system has the advantage of eliminating cumulative tolerance errors. Each dimension is

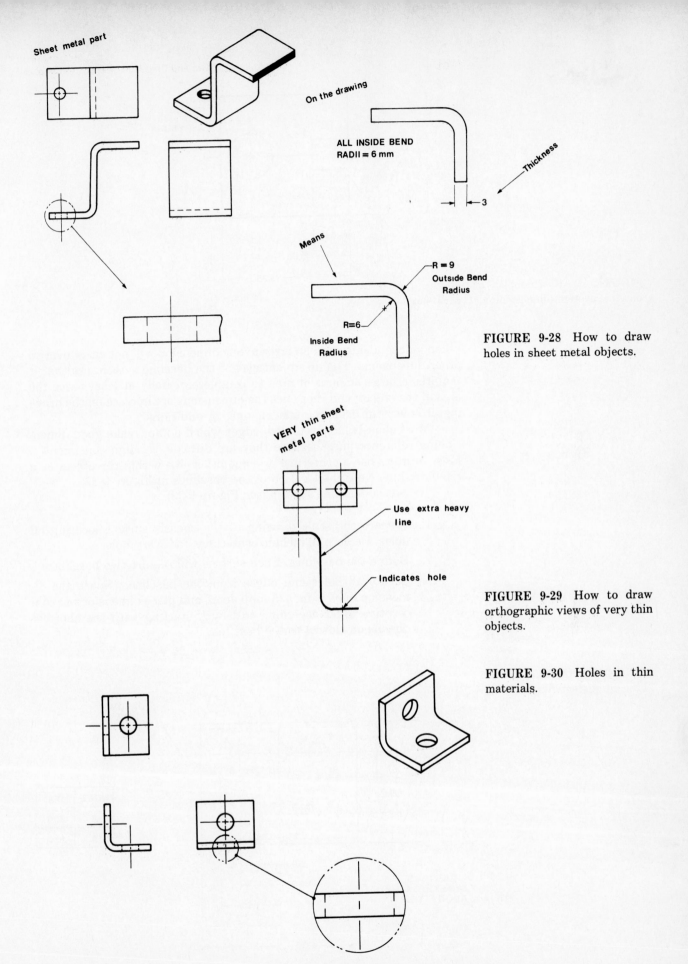

Sheet metal part

On the drawing

ALL INSIDE BEND
RADII = 6 mm

Thickness

3

Means

R = 9
Outside Bend
Radius

R=6

Inside Bend
Radius

FIGURE 9-28 How to draw holes in sheet metal objects.

VERY thin sheet metal parts

Use extra heavy line

Indicates hole

FIGURE 9-29 How to draw orthographic views of very thin objects.

FIGURE 9-30 Holes in thin materials.

193

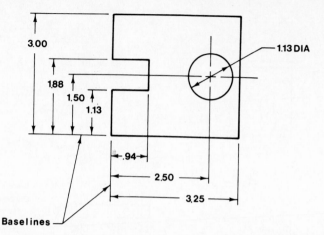

FIGURE 9-31 Example of baseline system dimensioning.

taken independently and an error in one dimension will not carry over to other dimensions. The disadvantages of the baseline system are that it requires a large amount of area to complete, usually at least twice the area of the object, and that once the dimensions are in place on the drawing, it is very difficult to make changes or additions.

Most chassis have rounded edges which do not make good dimensioning reference lines because they are difficult to align consistently. Most drafters use a centerline or some other line within the object as a reference line. Note the location of the baselines in Figure 9-32.

To use the baseline system (see Figure 9-33):

1. Prepare the scale drawing of the chassis surface locating all holes. Be sure to include centerlines for every hole.

2. Define the baselines. Each surface will require two baselines.

3. Locate all holes and edges from the baselines. Space the dimension lines ¼ inch (6 mm) apart and place dimensions in consecutive order according to length starting with the shortest dimension closest to the baseline.

FIGURE 9-32 Example of baseline system dimensioning.

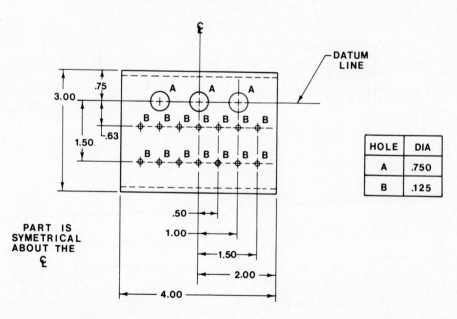

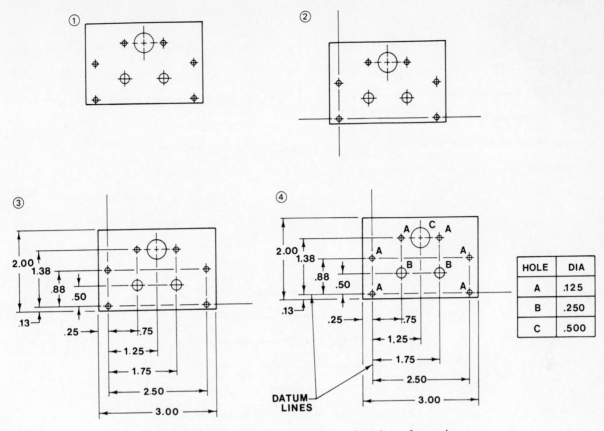

FIGURE 9-33   How to dimension a chassis surface using the baseline system.

4.   Define the hole sizes by assigning a letter to each hole size and then defining each letter in a chart, as shown in Figure 9-33. Place the letters on the drawing to the right and above the appropriate hole, if possible. Keep the letters close to the holes they define.

Figure 9-34 shows baseline dimensioning applied to five examples of repeated hole patterns. In example 1, the notation 5 × 06 means there are five holes of diameter 6 mm each. The parenthesis around the (25) dimensions means that the dimensions is a reference dimension and is not for manufacture or inspection. The notation 4 × 25 (= 100) dimensions means there are four spaces of 25 mm each. The " = 100" means that the total distance is 100 mm, but that this total distance is only for reference. Each 25 mm distance is to be considered, not the overall total.

In example 3, the 25.0 dimensions helps assure that the distance between the two holes is held to a tighter tolerance than the other holes. Example 4 shows how to dimension hole patterns with a hole pattern. Example 5 shows how baseline dimensioning is applied using polar coordinates.

The *coordinate system* is based on a 90° *x-y* coordinate system and is particularly well suited for use when programming numerically controlled machines. The coordinate system is easier to draw than the baseline system and is easier to change or correct but does require the reader

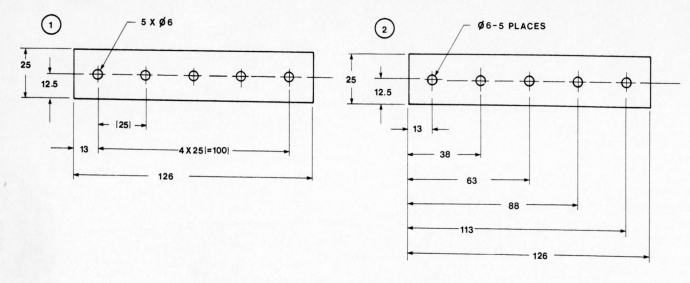

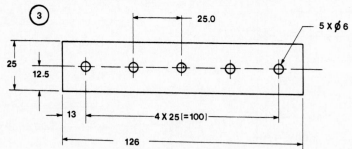

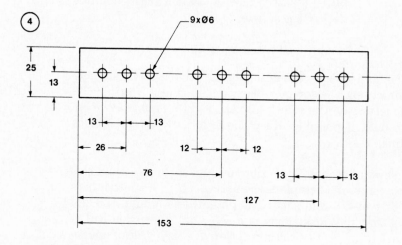

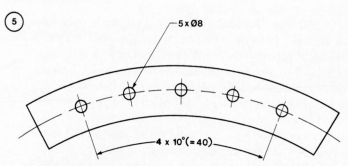

FIGURE 9-34   How to dimension hole patterns.

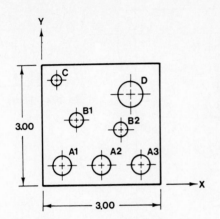

HOLE	X	Y	DIA
A1	0.50	0.50	.470
A2	1.50	0.50	.470
A3	2.50	0.50	.470
B1	0.88	1.62	.375
B2	2.00	1.38	.375
C	0.38	2.62	.250
D	2.25	2.25	.625

FIGURE 9-35   Example of coordinate system dimensioning.

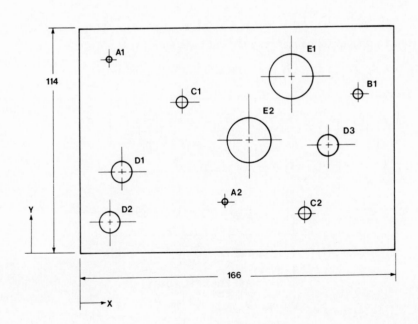

HOLE	DIAMETER	QTY
A	3.5	2
B	6.0	1
C	7.0	2
D	11.0	3
E	23.0	2

HOLE	FROM	X	Y
A1	X,Y	11	98
A2		76	25
B1		146	80
C1		54	76
C2		118	19
D1		22	41
D2		16	16
D3		131	54
E1		111	89
E2	X,Y	89	57

FIGURE 9-36   Rectangular coordinate system.

to look up each dimension. Figure 9-35 illustrates a chassis surface dimensioned using the coordinate system with inches.

Figure 9-36 shows a chassis surface dimensioned using the coordinate system with millimeter values. Figure 9-37 shows the coordinate system applied using polar dimensions.

To use the coordinate system (see Figure 9-38):

1.   Prepare a scale drawing of the chassis surface locating all holes. Be sure to include centerlines for every hole.

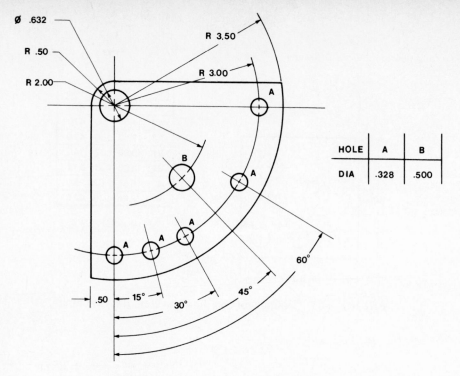

FIGURE 9-37 Polar coordinate system.

HOLE	A	B
DIA	.328	.500

FIGURE 9-38 How to dimension using the coordinate system.

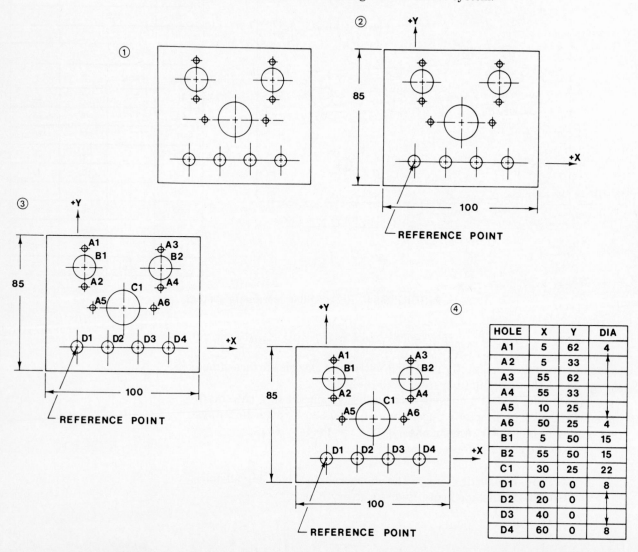

HOLE	X	Y	DIA
A1	5	62	4
A2	5	33	
A3	55	62	
A4	55	33	
A5	10	25	
A6	50	25	4
B1	5	50	15
B2	55	50	15
C1	30	25	22
D1	0	0	8
D2	20	0	
D3	40	0	
D4	60	0	8

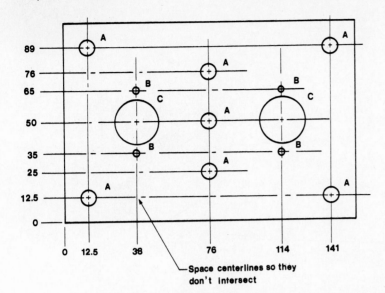

SIZE SYMBOL	A	B	C
HOLE DIA	8.5	4.0	28

Space centerlines so they
don't intersect

FIGURE 9-39 Rectangular dimension system without
dimension lines.

2. Define the reference point and then clearly label the positive x
    and y axes. Add the overall dimensions.

3. Label each hole using a letter and a number. Assign the same
    letter, but different consecutive numbers, to holes of equal diam-
    eter. In the example problem, the holes labeled A are all 4 mm
    in diameter; holes labeled B are 10 mm in diameter; and so on.

4. Prepare a chart which includes a listing, using the labels as-
    signed in step 3, of the distance that each hole is from the x and
    y axes and the hole's diameter.

Figure 9-39 shows a chassis surface dimensioned using a rectangular
dimensioning system without dimension lines. This system combines the
baseline and coordinate systems. The lower left corner in this example is
the (0,0) coordinate and the intersection of two perpendicular baselines.
All hole centerpoints are referenced to the (0,0) point. The diameter values
are listed in a chart as shown.

Problems

Figures P9-1 through P9-18 are dimensioned pictorial drawings of
objects. Draw three views (front, top, and right side) of each object as
assigned. The units are noted for each problem.

P9-1	Metric	**P9-7**	Metric	**P9-13**	Metric
P9-2	Metric	**P9-8**	Inches	**P9-14**	Metric
P9-3	Inches	**P9-9**	Metric	**P9-15**	Metric
P9-4	Inches	**P9-10**	Inches	**P9-16**	Metric
P9-5	Metric	**P9-11**	Metric	**P9-17**	Inches
P9-6	Metric	**P9-12**	Metric	**P9-18**	Inches

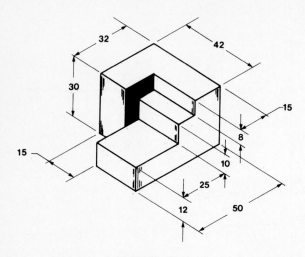

FIGURE P9-1

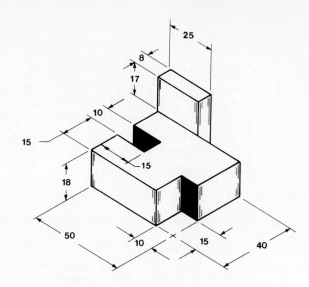

FIGURE P9-2

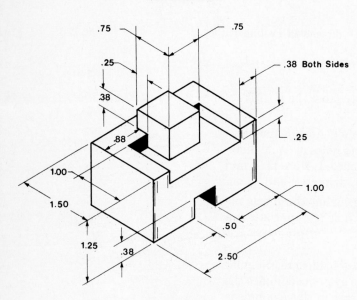

FIGURE P9-3

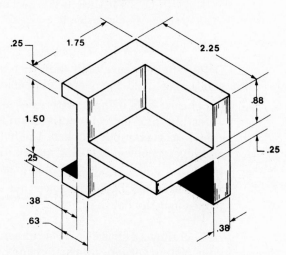

FIGURE P9-4

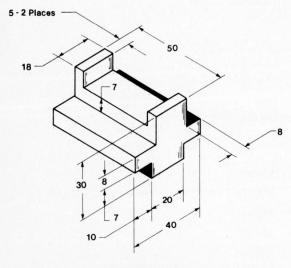

FIGURE P9-5

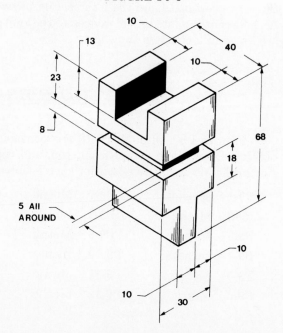

FIGURE P9-6

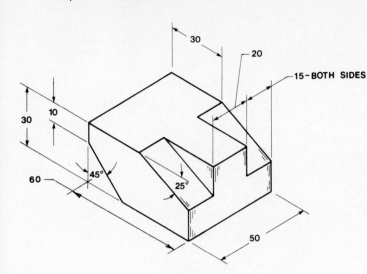

FIGURE P9-7

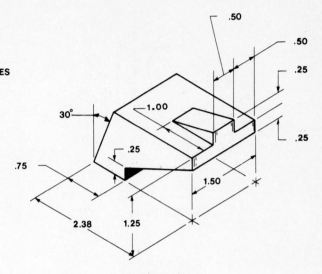

FIGURE P9-8

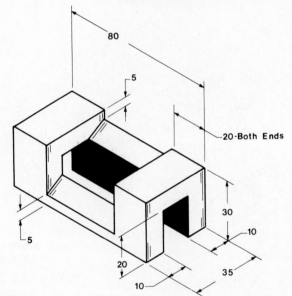

FIGURE P9-9

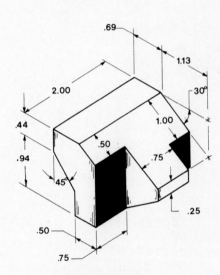

FIGURE P9-10

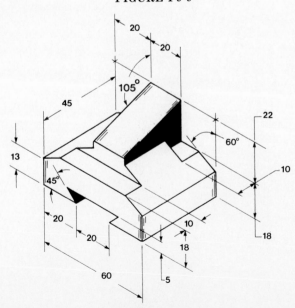

FIGURE P9-11

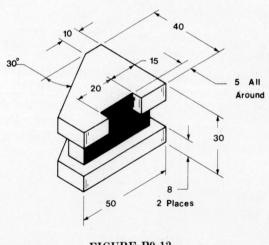

FIGURE P9-12

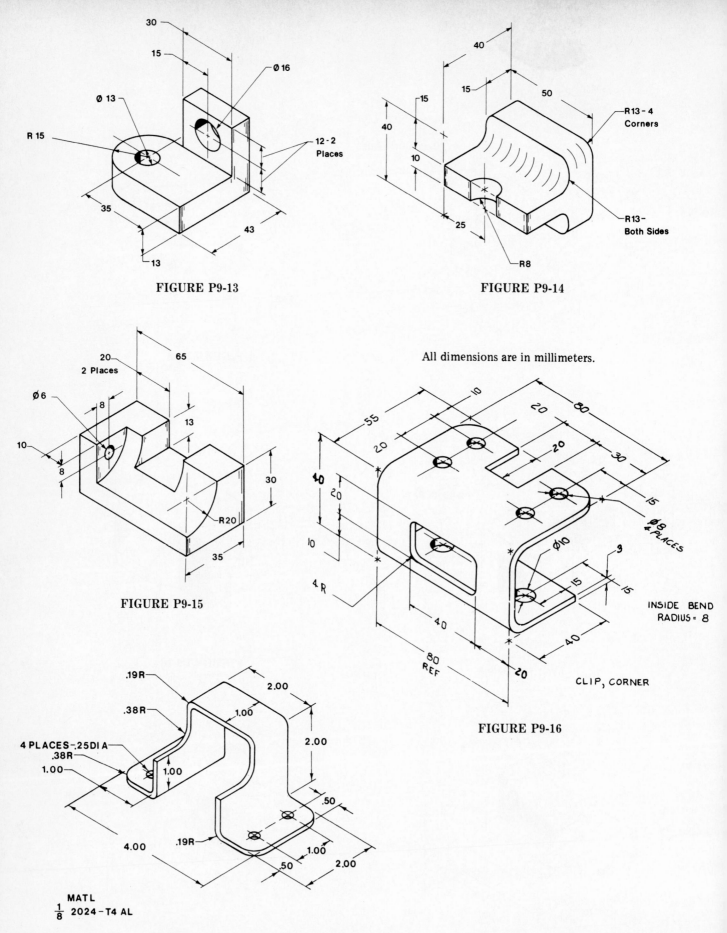

FIGURE P9-13

FIGURE P9-14

FIGURE P9-15

All dimensions are in millimeters.

CLIP, CORNER

FIGURE P9-16

INSIDE BEND
RADIUS = 8

MATL
$\frac{1}{8}$ 2024-T4 AL

FIGURE P9-17

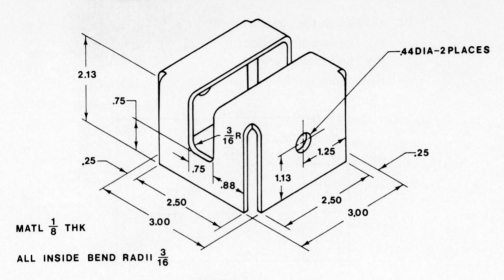

FIGURE P9-18

P9-19   Draw and dimension the flat pattern for the shapes defined in
Figure P9-19. All dimensions are in inches.

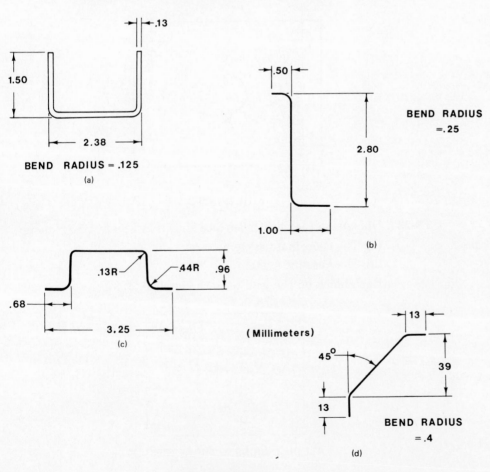

FIGURE P9-19

P9-20   Dimension the chassis surface shown in Figure P9-20 using:
        (a) The coordinate system
        (b) The baseline system
        Each block on the grid pattern equals 0.125 inch.

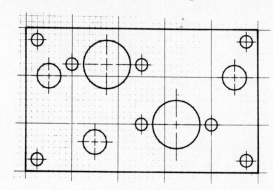

FIGURE P9-20

P9-21   Dimension the chassis surface shown in Figure P9-21 using:
        (a) The coordinate system
        (b) The baseline system
        Each block on the grid pattern equals 0.20 inch.

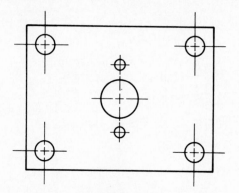

FIGURE P9-21

P9-22   Dimension the chassis surface shown in Figure P9-22 using:
        (a) The coordinate system
        (b) The baseline system
        Each block on the grid pattern equals 0.20 inch.

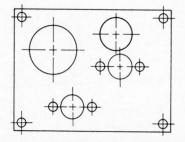

ALL DIMENSIONS TO THE NEAREST .10

FIGURE P9-22

P9-23   Draw and dimension a flat pattern for the chassis shape shown
in Figure P9-23. The holes are to be drilled before the material
is bent. Each block on the grid pattern equals 0.20 inch.

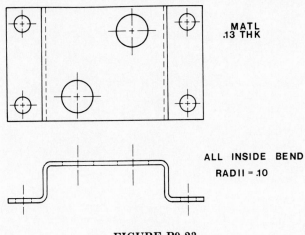

FIGURE P9-23

Redraw Figures P9-24 through P9-27 an add dimensions.

P9-24   Figures may be measured using either inches or millimeters. If
through   inches are used, measure to the nearest ¹⁄₃₂ or 0.03 in. If milli-
P9-27   meters are used, measure to the nearest millimeter.

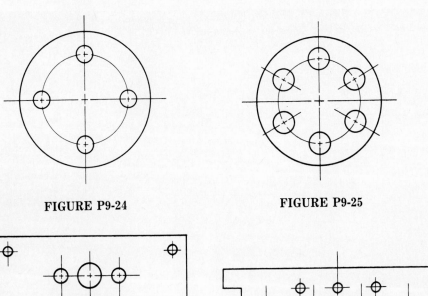

FIGURE P9-24   FIGURE P9-25

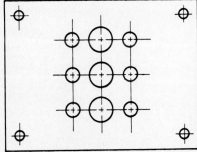   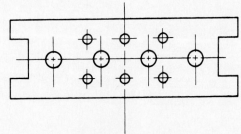

FIGURE P9-26   FIGURE P9-27

10

Connection Diagrams

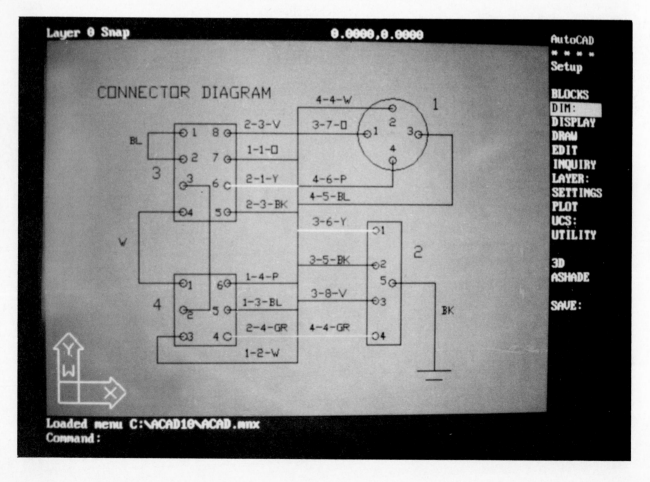

10-1 INTRODUCTION

Four different types of connection diagrams are described in this chapter. *Connection diagrams* are drawings that define how various components of a system are to be wired together. They are used most often in conjunction with assembly or maintenance instructions and as design layouts.

Each of the four types of connection diagrams discussed has advantages and disadvantages in its preparation and use. As you study how to draw the diagrams, try to become aware of these advantages and disadvantages so that you can learn which type is best suited for the requirements of each drawing.

10-2 POINT-TO-POINT DIAGRAMS

Point-to-point diagrams show the terminal connection location and routing path of every wire used in a system. Wires are not shown in bundles, nor do they use destination codes or tables. Each line drawn represents one wire which is graphically shown starting at one point and ending at another.

Point-to-point diagrams are very useful in design work because they enable the reader to follow directly the path of each wire. However, for large drawings, which contain many wires, point-to-point diagrams can become confusing and difficult to read accurately.

To draw a point-to-point diagram (see Figure 10-1):

1. Draw each component, locating it in the same position as that in which it is located in the installation. Label all components and terminals.

2. Draw in the wire paths. Wires that connect terminals on the same component are called *straps*.

FIGURE 10-1   How to prepare a point-to-point diagram.

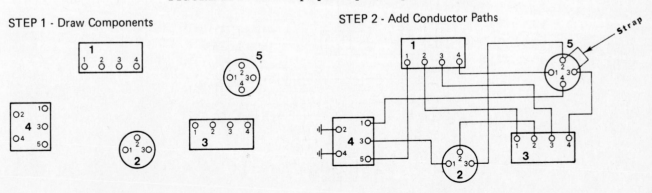

207

STEP 3 - Add Wire Colors

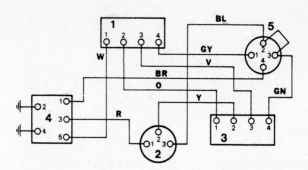

FIGURE 10-1 *(cont'd)*

3. If required, assign colors to each wire. A table of wire color codes is included in Appendix E.

Some drafters prefer to rearrange the terminal numbers so that they can make neater line patterns which are easier to follow. Figure 10-2 illustrates this technique. Note how, by changing the sequence of the terminal numbers, the diagram becomes much neater in appearance and much easier to follow.

FIGURE 10-2 Example of how rearranging the terminal numbers can be used to clarify a point-to-point diagram.

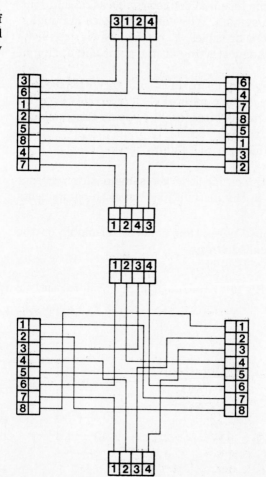

10-3 BASELINE DIAGRAMS

Baseline diagrams are drawings that feed all wires into one central line called a baseline. They present the wiring in a well-organized, easy-to-follow format, but have the disadvantage that they do not show components in their correct physical positions. Baseline drawings are particularly useful in presenting large, many-component diagrams in a size suitable for book use. They are used most often for maintenance and assembly manuals.

To draw a baseline diagram (Figure 10-3 illustrates):

1. Draw a light horizontal line across the center of the paper. This line is called the baseline.

2. Draw half of the components above the baseline and half below. Label all components and terminals.

3. Connect all used terminals to the baseline by drawing lines directly from the terminals to the baseline. If a direct path is not possible, make all directional changes at 90°. The baseline may not be bypassed. All wires must be connected to the baseline. (Straps need not be connected to the baseline.)

4. Label all wires using a destination code.The destination code

FIGURE 10-3   How to prepare a baseline diagram.

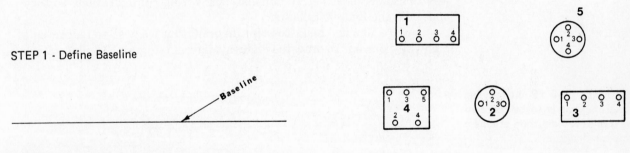

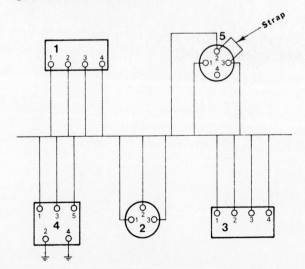

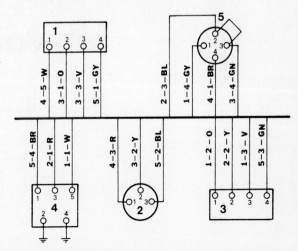

consists of letters and numbers that identify the component number and terminal number to which the wire is to be attached, and the color of the wire. See Figure 10-4 for a further explanation of the destination code.

5.   Darken in the baseline, making it a very heavy, black line.

FIGURE 10-4   Destination code.

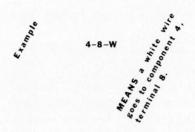

10-4   HIGHWAY DIAGRAMS

Highway diagrams combine groups of wires running along similar paths into bundles called *highways*. The components are located in the same relative positions as they would be located in the actual component setup. Of the different types of component diagrams shown in this chapter, highway diagrams most nearly duplicate the wiring configurations as they appear in the final installation.

Figure 10-5 is a point-to-point diagram that we wish to redraw as a highway diagram. To draw the highway diagram (Figure 10-6 illustrates):

FIGURE   10-5   Point-to-point diagram that is to be redrawn as a highway diagram and as a lineless diagram.

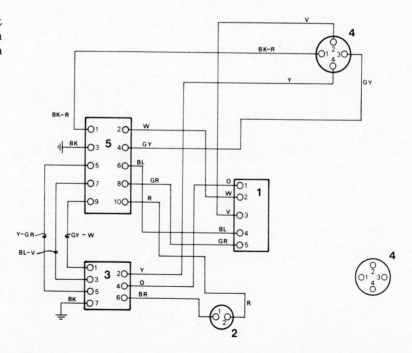

1. Draw the components, locating them in the same positions as those they occupy in the final installation. Identify all components and terminals.

2. Study the drawing and with light construction lines, identify all possible areas where wires may be bundled. These bundles are called highways.

STEP 1 - Arrange Components

FIGURE 10-6   How to prepare a highway diagram.

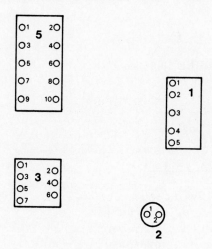

STEP 2 - Plan Highways

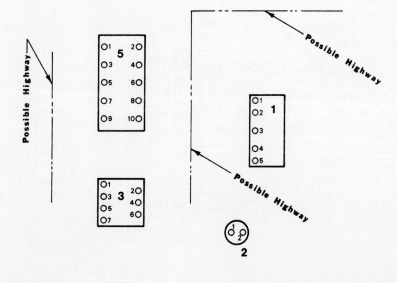

3. Draw very light lines from the appropriate terminals to the highways. Where an individual wire joins the highway, use either an arc or a line slanted at 45°, as shown in Figure 10-7. The direction of the arc or slanted line should be in the direction the wire is headed. For example, the arc showing the intersection of the wire from terminal 10 of component 5 to the highway turns down as the wire is headed for component 2, whereas the wire from terminal 3 of component 1 turns up as the wire is headed for component 4. *Note that unlike baseline diagrams, highways may be bypassed. The wire from terminal 6 of component 3, for example, may be drawn directly to terminal 1 of component 2. It need not be included as part of a highway.*

4. Label each wire with a destination code: component number–terminal number–wire color. Ground wires need only be labeled with their colors and the symbol for ground.

FIGURE 10-6 *(cont'd)*

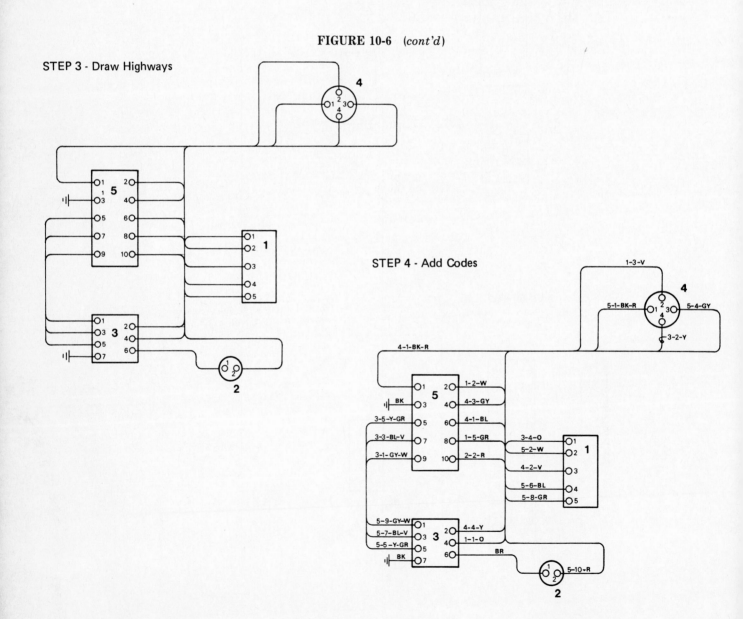

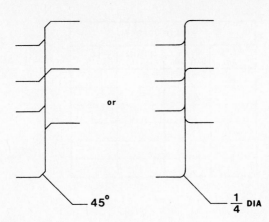

FIGURE 10-7 Wire turns and intersections on a highway drawing may be drawn as either an arc or as a line slanted at 45°.

10-5   LINELESS DRAWINGS

Lineless drawings are component drawings that do not show any wiring paths but instead define the wiring paths by using a wiring table. Lineless diagrams are particularly useful for clarifying large complex drawings where the number of wires is so great that several sheets of drawings would be required to include the entire pattern. The disadvantage of lineless diagrams is that because they omit all line patterns, they present an incomplete representation of what the final wiring setup will look like.

Figure 10-5 is a point-to-point diagram that we wish to redraw as a lineless diagram. To draw the lineless diagram (Figure 10-8 illustrates):

1. Draw the components of the system. If possible, locate the components on the drawing in positions that approximate their actual physical locations, and number the terminals. Each component must be labeled.

FIGURE 10-8   Example of a lineless diagram.

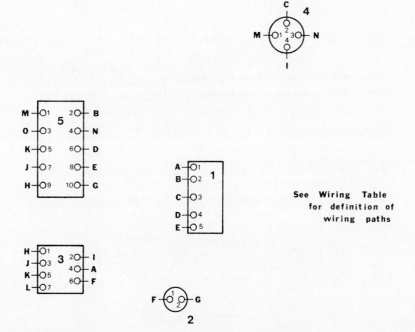

WIRING TABLE			
WIRE I.D.	WIRE COLOR	FROM	TO
A	O	1–1	3–4
B	W	1–2	5–2
C	V	1–3	4–2
D	BL	1–4	5–6
E	GR	1–5	5–8
F	BR	2–1	3–6
G	R	2–2	5–10
H	GY–W	3–1	5–9
I	Y	3–2	4–4
J	BL–V	3–3	5–7
K	Y–GR	3–5	5–5
L	BK	3–7	GRD
M	BK–R	4–1	5–1
N	GY	4–3	5–4
O	BK	5–3	GRD

WIRING TABLE			
WIRE I.D.	WIRE COLOR	FROM	TO
A	G	1–1	3–2

Sample Callout

FIGURE 10-9 Example of a wiring table that would accompany a lineless diagram.

2. Prepare a wiring chart such as the one shown in Figure 10-9. The chart should include wire identification, where the wire starts (component and terminal), and where the wire ends (component and terminal). The chart may also include wire size and wire color. In this example, the wires were identified by letters and the components by numbers. Component names could also have been used.

3. Draw a short line from each terminal used and label it with the appropriate wire identification letter.

10-6 HARNESS ASSEMBLY DRAWINGS

A *harness assembly drawing* defines the pattern for wiring in an electrical assembly and how the wires are to be bundled together. Figure 10-10 is an example of a harness assembly drawing.

Wires are bound together in groups called harnesses, as shown in Figure 10-11. Wiring harnesses are made outside the assembly. This makes it easier to bundle the wires (the components are not in the way) and also helps to prevent damage to the assembly.

FIGURE 10-10 Sample harness assembly diagram.

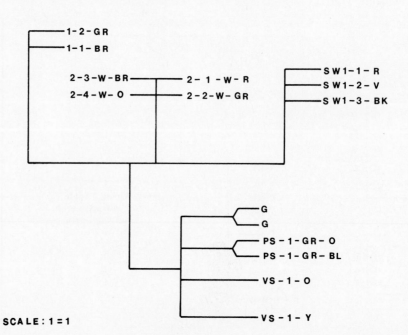

SCALE: 1 = 1

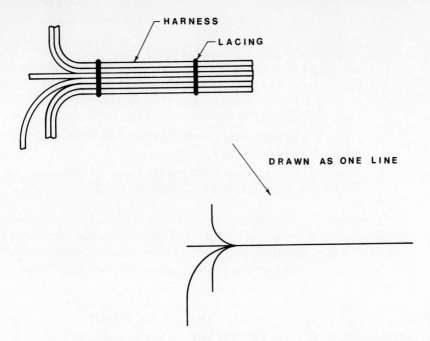

FIGURE 10-11 Harnesses.

Harness assembly drawings are drawn at a 1:1 scale: in other words, at their actual size. This is done so that the drawing can be used to help construct the harness.

To create a harness assembly drawing (see Figure 10-12):

1. Create a layout drawing, using a 1:1 scale, of all components and connections in the assembly. Draw all components using their actual size and locate them using the actual distance from each other. Identify all components.

2. Locate and identify all (even those not used) terminals on all components [see Figure 10-12(a)].

FIGURE 10-12   How to create a harness assembly diagram.

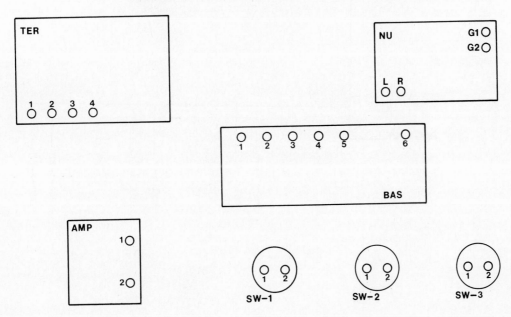

(a)

215

3. Add the wire paths as needed [see Figure 10-12(b)]. Wire paths are added using a color different than that used to define the components. The wire paths are then combined to form bundles. Several attempts are usually required before a final arrangement is obtained.

4. Identify all leads using the destination code explanation in Figure 10-4. Define the wire colors.

5. Create a table as shown. All lengths may be measured off the harness drawing layout [see Figure 10-12(c)].

There are many different pieces of information that could be listed in a harness table, including lacing termination distance, American wire gage (AWG) number, insulation type, type of wire, and lead termination requirements. The table presented here gives only the most basic information.

FIGURE 10-12 *(cont'd)*

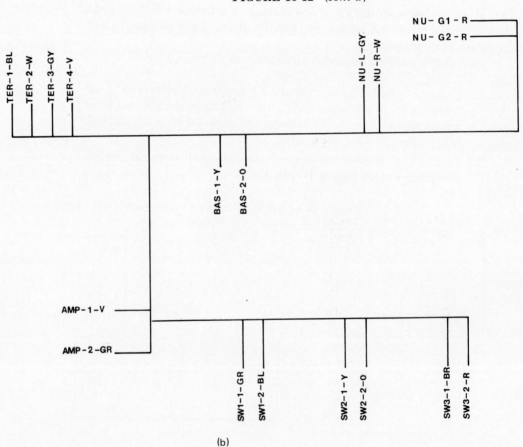

(b)

FROM LOCATION	LEAD TERMINATION (STRIPPED)	COLOR OF INSULATION	LEAD LENGTH (INCHES)	TO LOCATION	LEAD TERMINATION (STRIPPED)
TER-1	$\frac{3}{4}$	BL	$8\frac{3}{16}$	SW1-2	$\frac{3}{4}$
TER-2	$\frac{3}{4}$	W	$6\frac{13}{16}$	NU-R	$\frac{3}{4}$
TER-3	$\frac{3}{4}$	GY	$6\frac{1}{8}$	NU-L	$\frac{3}{4}$
TER-4	$\frac{3}{4}$	V	$10\frac{9}{16}$	NU-G2	$\frac{1}{2}$
AMP-1	$\frac{3}{4}$	O	$5\frac{3}{16}$	BAS-2	$\frac{1}{2}$
AMP-2	$\frac{1}{2}$	GR	$2\frac{5}{16}$	SW1-1	$\frac{3}{4}$
SW1-1	$\frac{3}{4}$	GR	$2\frac{5}{16}$	AMP-2	$\frac{1}{2}$
SW1-2	$\frac{3}{4}$	BL	$8\frac{3}{16}$	TER-1	$\frac{3}{4}$
SW2-1	$\frac{1}{2}$	Y	$8\frac{5}{8}$	BAS-1	$\frac{1}{2}$
SW2-2	$\frac{1}{2}$	BR	2	SW3-1	$\frac{3}{4}$
SW3-1	$\frac{3}{4}$	BR	2	SW2-2	$\frac{1}{2}$
SW3-2	$\frac{3}{4}$	R	$18\frac{1}{4}$	NU-G1	$\frac{1}{2}$
NU-L	$\frac{3}{4}$	GY	$6\frac{1}{8}$	TER-3	$\frac{3}{4}$
NU-R	$\frac{3}{4}$	W	$6\frac{13}{16}$	TER-2	$\frac{3}{4}$
NU-G1	$\frac{1}{2}$	R	$18\frac{1}{4}$	SW3-2	$\frac{3}{4}$
NU-G2	$\frac{1}{2}$	V	$10\frac{9}{16}$	TER-4	$\frac{3}{4}$
BAS-1	$\frac{1}{2}$	Y	$8\frac{5}{8}$	SW2-1	$\frac{1}{2}$
BAS-2	$\frac{1}{2}$	O	$5\frac{3}{16}$	AMP-1	$\frac{3}{4}$

(c)

FIGURE 10-12 (*cont'd*)

Problems

P10-1 Given the point-to-point diagram in Figure P10-1, redraw the diagram as:

(a) A highway diagram

(b) A baseline diagram

(c) A lineless diagram

(d) A harness assembly drawing

In preparing the harness assembly drawing, assume that all components are drawn and located at a 1:1 scale.

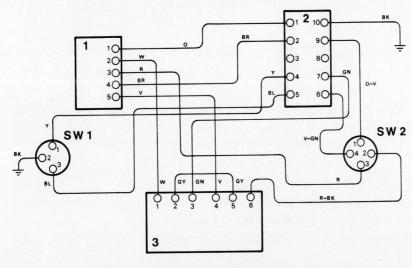

FIGURE P10-1

P10-2   Figure P10-2 is an illustration of a home stereo system which includes two speakers, an amplifier, and a turntable. Redraw the system as:

(a) A highway diagram

(b) A baseline diagram

(c) A lineless diagram

(d) A point-to-point diagram

Assign component numbers, terminal numbers, and wire colors as needed. Wire color abbreviations are given in Appendix E.

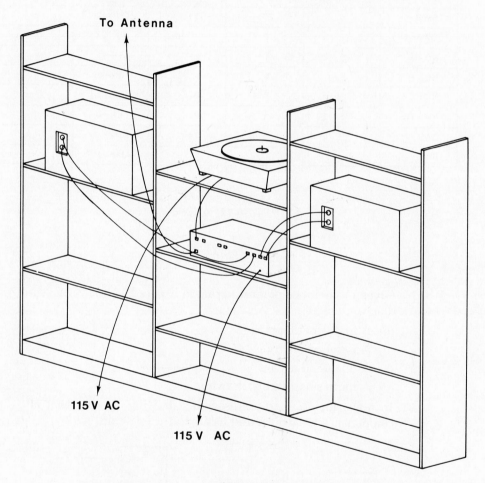

FIGURE P10-2

P10-3   Your company is working on a new electronic system. A prototype has been set up in the lab and you have been asked to study the prototype setup (pictured in Figure P10-3) and prepare a wiring diagram. Redraw the system as:

(a) A highway diagram

(b) A baseline diagram

(c) A lineless diagram

(d) A point-to-point diagram

(e) A harness assembly drawing

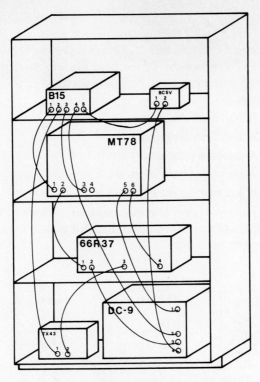

FIGURE P10-3

P10-4   Figure P10-4 is an engineer's design sketch. Redraw the sketch
as:

(a) A highway diagram

(b) A baseline diagram

(c) A lineless diagram

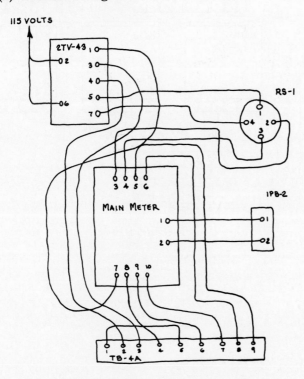

DESIGN SKETCH - DIPLEX SYSTEM

FIGURE P10-4

P10-5 Figures P10-5 through P10-8 are representative of the types of
through design sketches for which drafters are often asked to create
P10-8 appropriate drawings. For each problem, create one of the following:

(a) A highway diagram

(b) A baseline diagram

(c) A lineless diagram

(d) A point-to-point diagram

(e) A harness assembly drawing

When preparing a harness assembly drawing, assume that all
components are shown and located at a 1:1 scale. Assign wire
colors as necessary.

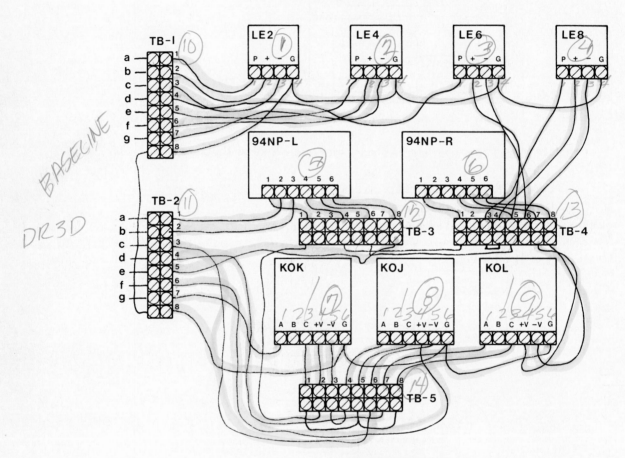

FIGURE P10-5

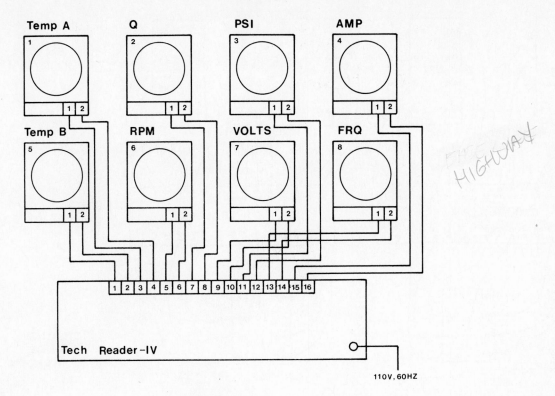

Temp A   Q   PSI   AMP

Temp B   RPM   VOLTS   FRQ

Tech   Reader -IV

110V, 60HZ

FIGURE P10-6

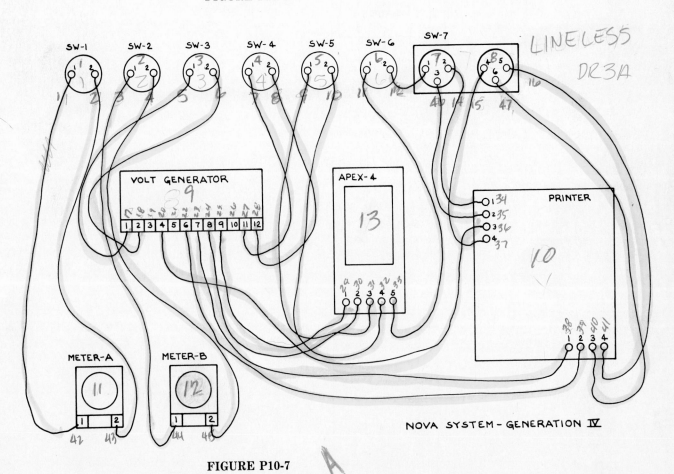

SW-1   SW-2   SW-3   SW-4   SW-5   SW-6   SW-7

LINELESS
DR3A

VOLT GENERATOR

APEX-4

PRINTER

METER-A   METER-B

NOVA SYSTEM- GENERATION IV

FIGURE P10-7

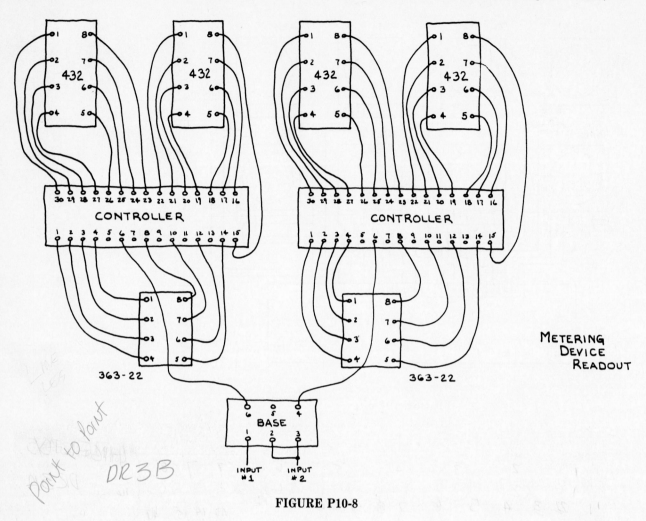

FIGURE P10-8

11

Residential Electrical Wiring

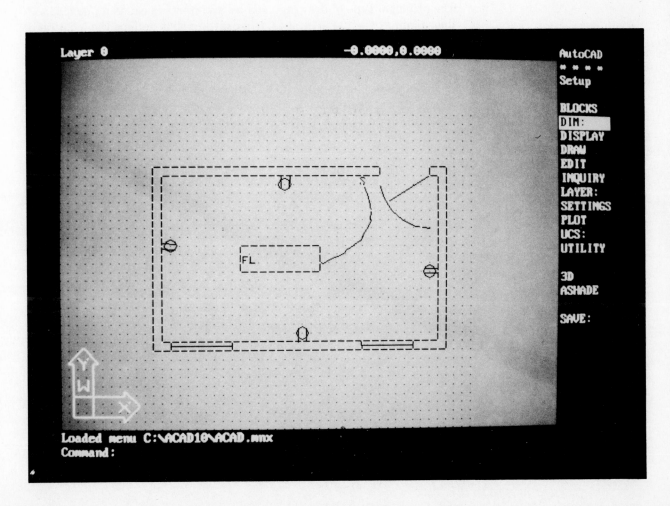

11-1 INTRODUCTION

This chapter deals with residential electrical wiring drawings. We explain basic fundamentals of architectural drawing, define and show the application of the graphic symbols used on residential wiring drawings, and cover some of the basic design concepts required for residential wiring.

The information presented is in agreement with the standards set forth in the *National Electrical Code®* published by the National Fire Protection Association, 470 Atlantic Ave., Boston, MA 02110. Interested students may purchase copies of the Code by writing to the NFPA at the address above.

11-2 BASIC ARCHITECTURAL DRAFTING

The principal type of drawing used by architects is called a *floor plan*. A floor plan, as the name implies, is a drawing that defines the location of the various rooms, windows, doors, stairs, closets, hallways, and so on, for a given residence. Figure 11-1 shows a floor plan together with an elevation drawing of a small one-bedroom house.

The floor plan in Figure 11-1 has been labeled so that you can learn how to interpret architectural drawings. Note how doors, windows, and so on, are represented. Drafters usually create a symbol library such as

FIGURE 11-1 Floor plan and elevation drawing of a small house. (The elevation drawing was prepared by Raymond Collard and the house design was created by Arthur Nelson, both of Wentworth Institute of Technology.)

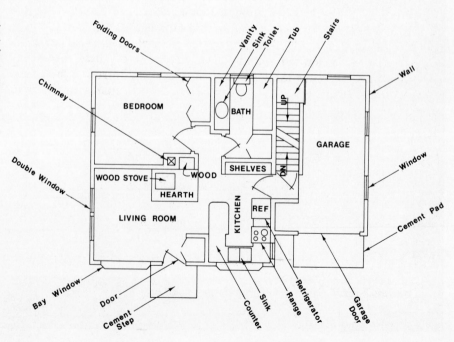

FIGURE 11-1   *(cont'd)*

outlined in Section 4-2 for those symbols used most often. These symbols may include doors, windows, kitchen, and bathroom fixtures as well as some large home furnishings.

The standard scale used for residential drawings is ¼ inch = 1 foot; every ¼ inch on the drawing equals 1 foot on the house. Figure 11-2 illustrates a ¼-inch scale together with some sample measurements. The ¼ inch to the right of the 0 mark on the sample scale is graduated into 12 equal divisions so that measurements in inches can be made. Some scale manufacturers combine the ¼-inch scale, which reads from right to left, with a ⅛-inch scale, which reads left to right. The smaller numbers included in Figure 11-2 refer to values on a ⅛-inch scale.

FIGURE 11-2   A ¼-inch scale with some sample readings.

Readings are based on a scale of $\frac{1''}{4}=1'$

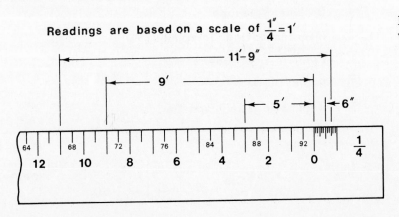

11-3 ELECTRICAL SYMBOLS

Symbols are used on residential electrical wiring drawings to specify the type and location of the switches or outlets required. Figure 11-3 illustrates the most commonly used symbols and their meaning. The dimensions given are only approximations, as no national standard sizes have been agreed upon.

Figure 11-4 illustrates how these symbols may be added to a floor plan to create an electrical wiring drawing. It is suggested that these symbols be redrawn and filed to create a symbol library as outlined in

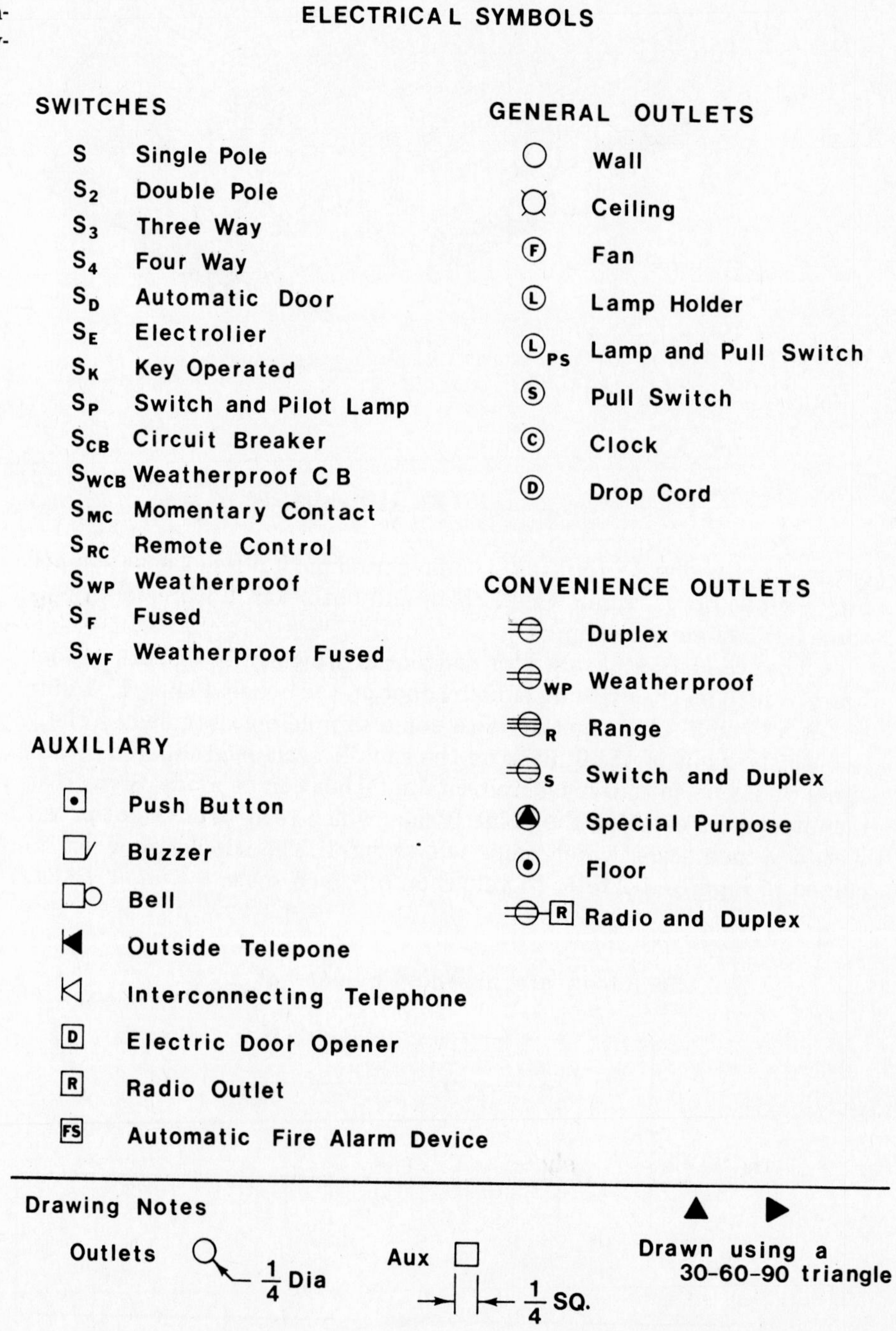

FIGURE 11-3 Electrical symbols for residential wiring drawings.

ELECTRICAL SYMBOLS

SWITCHES

S	Single Pole
S_2	Double Pole
S_3	Three Way
S_4	Four Way
S_D	Automatic Door
S_E	Electrolier
S_K	Key Operated
S_P	Switch and Pilot Lamp
S_{CB}	Circuit Breaker
S_{WCB}	Weatherproof C B
S_{MC}	Momentary Contact
S_{RC}	Remote Control
S_{WP}	Weatherproof
S_F	Fused
S_{WF}	Weatherproof Fused

AUXILIARY

Push Button
Buzzer
Bell
Outside Telepone
Interconnecting Telephone
Electric Door Opener
Radio Outlet
Automatic Fire Alarm Device

GENERAL OUTLETS

Wall
Ceiling
Fan
Lamp Holder
Lamp and Pull Switch
Pull Switch
Clock
Drop Cord

CONVENIENCE OUTLETS

Duplex
Weatherproof
Range
Switch and Duplex
Special Purpose
Floor
Radio and Duplex

Drawing Notes

Outlets $\frac{1}{4}$ Dia   Aux $\frac{1}{4}$ SQ.   ▲ ▶ Drawn using a 30–60–90 triangle

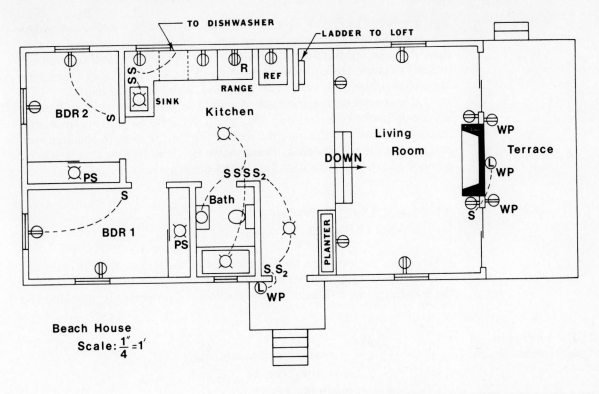

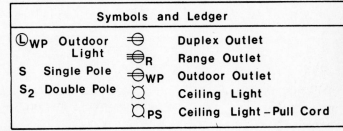

FIGURE 11-4   Example of a residential wiring drawing.
The floor plan is for a summer beach house.

Section 4-2. In some cases, the electrical wiring drawing is created by making a drawing of the floor plan and then drawing on the appropriate symbols using another color, such as red or yellow.

Wiring paths are drawn as hidden lines (a pattern of evenly spaced dashes). They are drawn using irregular curves so that they are easily distinguishable from the floor plan lines. This is done by using a SPLINE function or a freehand, random points line. Note that the wire paths are specified only between switches and the fixtures they activate. It is not necessary to show all the wiring paths within a residence.

Wiring paths are omitted for two reasons. To show all the wiring would create a very cluttered, difficult-to-follow drawing which could lead to interpretation errors. It is also not necessary to show all the paths. Electricians need only be shown the location and type of fixture required and they will be responsible to ensure that the fixture is wired properly in accordance with the NEC and any local variances.

Every residential electrical wiring drawing you create should include a symbol ledger which illustrates each symbol used on the drawing and defines its meaning. A sample symbol ledger is located on the drawings presented in Figure 11-4. The ledger may also be used to define any new symbols needed. Sometimes special equipment, such as smoke detectors, must be specified on a drawing and no standard symbol has been defined to cover the equipment. In such cases, a drafter may make up a symbol *provided* that it is clearly defined in the drawing ledger.

If a special or specific type of switch or outlet is required, it must also be defined on the drawing. This is usually done by printing the manufacturer's name and the part number of the fixture next to the appropriate symbol on the drawing. It may also be done by creating a special symbol and defining its meaning in the drawing ledger.

11-4 LOCATING SWITCHES AND OUTLETS

When choosing locations for electrical switches and outlets, it is important to consider how the switches and outlets will be used. For example, the switch shown in Figure 11-5 is not located properly. It is behind the

FIGURE 11-5 Examples of good and poor electrical design.

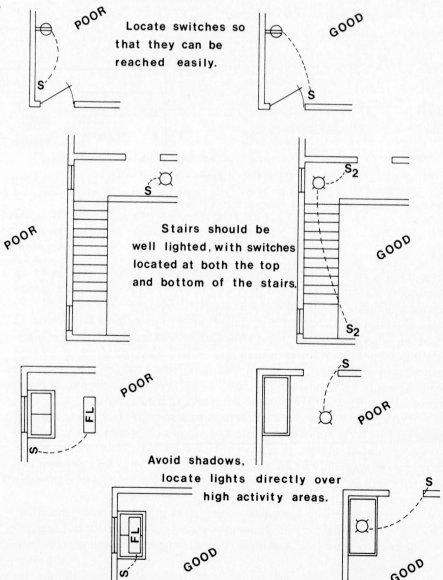

door and would require a person entering the room to start to close the door in order to reach the switch. This is unacceptable, particularly if the room is dark.

Another error illustrated in Figure 11-5 concerns the stairs. There is no light located near or over the stairs, meaning that the stairs cannot be seen clearly at night.

The kitchen, shown in Figure 11-5 is also not lighted properly, even though the large fluorescent fixture would seem to be able to generate more than enough light. The problem here is one of shadows. A person working at the sink would have the light at his or her back, causing a shadow to fall over the sink. Even though the room is well lighted, the sink would be in darkness, making it inconvenient to use.

Shadows can also cause problems in the bathroom. In Figure 11-5 there is a light over the sink, but none over the tub. If a shower curtain were added (this would not necessarily appear on the floor plan) with the lighting setup shown, the light would be blocked out by the curtain and the person showering would be in greatly reduced light.

It is good practice, when preparing your drawing, to pretend that you are actually in the house on which you are working. Asssume that it is dark, and mentally walk through the house checking that all switches and outlets that you have specified are located conveniently. For example, if you are working on a two-story house, and you are on the first floor, you will want a light over the stairs, a switch on the first floor to activate the light, and another switch at the top of the stairs to deactivate the light.

Lights can be arranged with two, three, or four separate activating switches. The symbols for double-pole, three-way, and four-way switches are shown in Figure 11-6.

FIGURE 11-6   How to indicate double-pole, three-way, and four-way switches on a drawing.

A single pole switch which is wired to a duplex outlet; the switch activates or deactivates the outlet.

Two double pole switches wired to a single ceiling outlet; either switch can activate or deactivate the outlet.

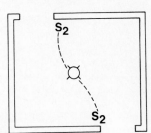

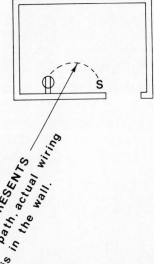

Three three-way switches all wired to a single ceiling outlet; any of three switches can be used to activate or deactive the outlet.

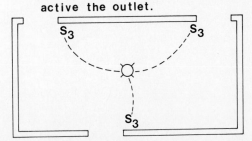

To help you locate switches properly, the following checklist has been prepared. Check your drawings against the list to help prevent embarrassing errors.

1. When entering a room, is a light switch easily accessible?
2. For rooms with two or more entrances, are light switches located at each entrance?
3. Are all outside entrances lighted?
4. Are all hallways and stairs lighted with switches located at each end?
5. Are all work areas (sinks, counters, etc.) lighted so as to prevent excessive shadows?
6. Are all special electrical requirements (range, dishwasher, dryer, air conditioners, etc.) taken into account?
7. Are all outside fixtures waterproof?

11-5 ELECTRICAL TERMINOLOGY

Several terms commonly used in residential electrical wiring are listed below. Each is defined and explained. Figure 11-7 illustrates some of the terms.

Branch Circuit   A power-conducting wiring path that leads from the service panel to a specific area of the house and back to the service panel.

FIGURE 11-7   Residential wiring terms.

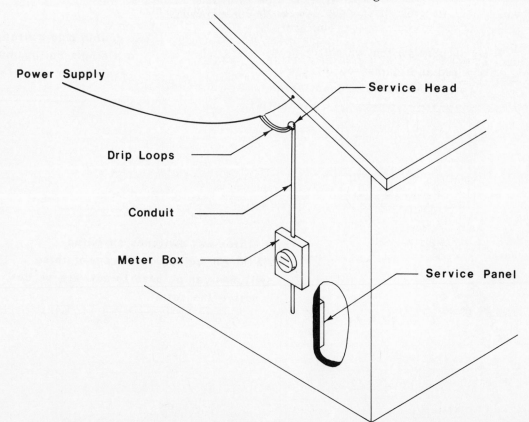

Some pieces of equipment used in residences draw so much electricity that a special individual branch circuit is created just for them. Electric ranges, dryers, refrigerators, air conditioners, and heaters are some of the equipment that require an individual branch circuit.

Cable   Electrical wiring that is covered with a protective cover or insulating material. Romex cable, copper wires covered with heavy paper and material for use in dry locations and covered with plastic for use in damp locations, is the type most often used for houses. Armored cable, BX, is sometimes used.

Conduit   A thin-gaged metal pipe or tube used to protect cable from damage.

Drip Loop   A way to suspend the service wires so as to prevent water from running down the wire into the meter or service panel.

Mast   Conduit that extends above the roof of a house and is used to receive the service wires. Masts are not always used.

Meter Box   A gage installed by the local power company to keep track of the amount of electricity used.

Service Head   A metal cover placed on top of conduit to prevent water from entering the meter or service panel.

Service Panel   A metal box that contains the main disconnect switch and circuit breakers. It also contains all connections between the main incoming power and the branch circuits.

Service Wires   The power lines that bring electricity from the power company's lines to the house. The power supplied is usually 240/120-volt single-phase 60-hertz ac.

15-ampere Line, 20-ampere Line, 240-volt Line   These terms refer to different types of branch circuits. A 15-ampere line is an average line that would service a bedroom, bathroom, or similar room. A 20-ampere line would be used for circuits that will probably get heavy use, such as in a kitchen where many different appliances are used. A 240-volt line is a special, very heavy duty branch circuit which is designed to carry large amounts of current. An electric hot water heater, electric range, or electric dryer would all require 240-volt lines.

Problems

P11-1   Figure P11-1 shows a floor plan of a bedroom, including a closet. Redraw the floor plan using a scale of ¼ inch = 1 foot and add, using the appropriate electrical symbols: three duplex outlets, one of which is wired to a switch; a pull switch in the closet; and a telephone jack.

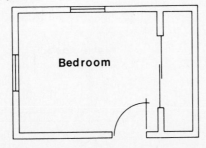

FIGURE P11-1   Floor plan of a bedroom. Scale: ⅛ inch = 1 foot.

P11-2   Figure P11-2 shows the floor plan for a kitchen. Redraw the floor plan using a scale of ¼ inch = 1 foot and add, using the appropriate electrical symbols, each of the following:

(a) A range outlet

(b) A dishwasher outlet

(c) A refrigerator outlet

(d) An overhead light with activating switches at each door

(e) A light over the sink, with an activating switch

(f) A light and a fan over the range, with an activating switch for each

(g) Four duplex outlets located over the counters

(h) A wall outlet for a clock

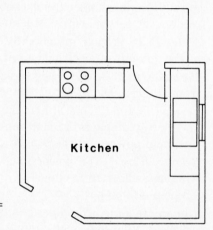

FIGURE P11-2   Floor plan of a kitchen. Scale: ⅛ inch = 1 foot.

P11-3   Figure P11-3 shows the floor plans for a two-story house. Redraw the floor plans using a scale of ¼ inch = 1 foot, and use the plans as a basis to create a set of residential electrical drawings.

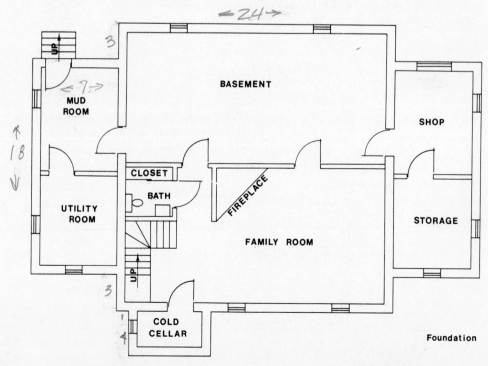

FIGURE P11-3   Floor plan of a two-story house. Scale: ⅛ inch = 1 foot.

Think carefully about the locations of all switches, outlets, and so on, before you add them to the drawings. Redraw the floor plans, as assigned by your instructor, as follows:

(a) First floor

(b) Second floor

(c) Foundation

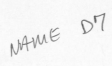

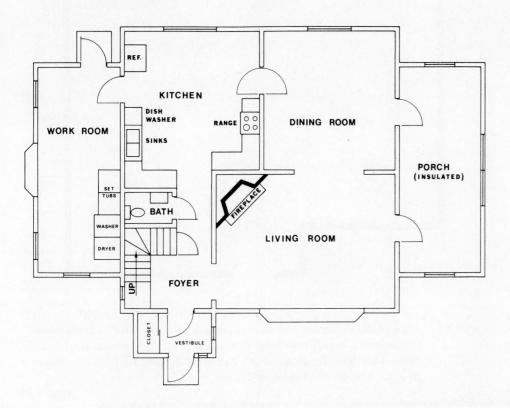

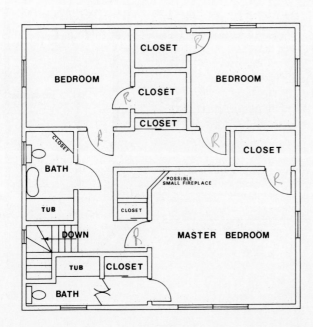

P11-4   Figure P11-4 shows the floor plan for a loft-type apartment. It is
         drawn at a scale of ⅛ inch = 1 foot. The owner has asked you to
         prepare an electrical plan based on the information given. Use a
         scale of ¼ inch = 1 foot.

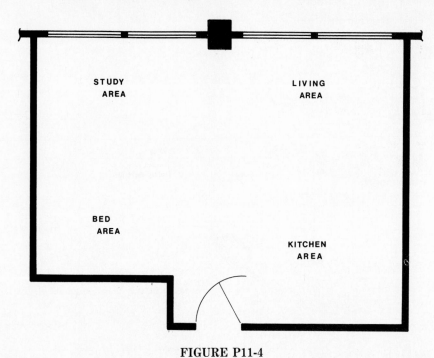

FIGURE P11-4

P11-5   Figure P11-5 shows the floor plan for a house that is built into
         the side of a hill. It looks out over the ocean. Prepare an electrical
         plan. The drawing is done at ⅛ inch = 1 foot. Redraw the floor
         plan at a scale of ¼ inch = 1 foot.

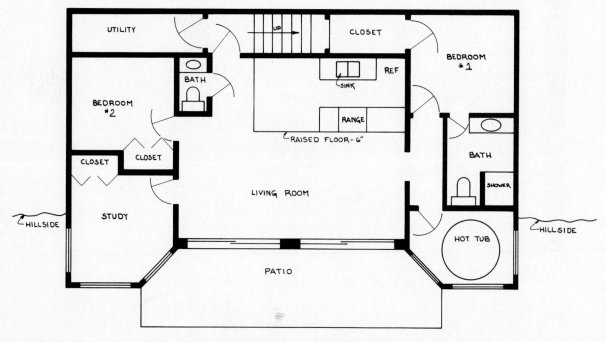

FIGURE P11-5

12

Industrial Wiring Diagrams

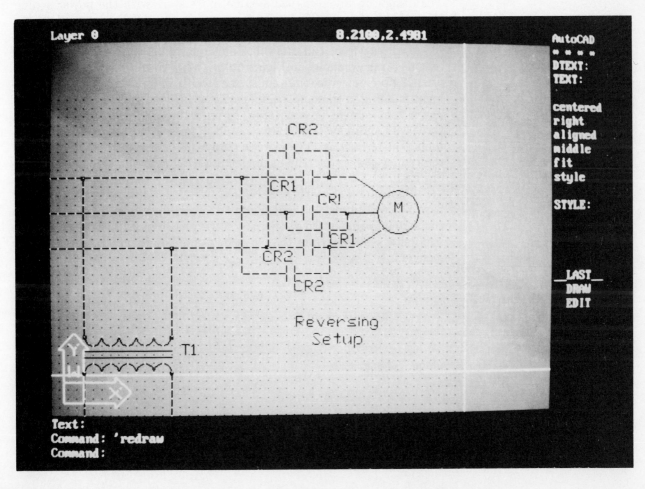

12-1 INTRODUCTION

In this chapter we explain how to draw some of the most widely used types of industrial wiring diagrams, including one-line diagrams, ladder diagrams, and raiser diagrams. Also presented are some of the basic design and electrical principles which these diagrams represent. The student interested in more complete coverage of the subject is referred to:

Charles W. Snow, *Electrical Drafting and Design.* Englewood Cliffs, N.J.: Prentice-Hall, Inc., 1976.

The symbols and line techniques required for each type of diagram are presented together with the text material.

12-2 BASIC POWER SYSTEMS

The design of any electrical system must start with the power source. What kind of power is available, and how much? For example, most houses are supplied by the local power company with power that is single-phase, three-wired, 120/240 volts, at 60 hertz. This would be expressed on a drawing as shown in Figure 12-1.

If more power is required, such as would be needed for a small apartment building, a three-phase, four-wire wye system could be used. If the power requirement is very large, as for a machine shop, a three-phase, three-wire, delta system could be used. Figure 12-2 illustrates these systems. The terms *wye* (Y) and *delta* (Δ) refer to the shape of windings used to produce the power. The basic patterns and symbols are shown in Figure 12-3.

Figure 12-4 shows how a single phase, three-wire system can be used to produce different voltages. For a house, these connections are made in the service panel. Also shown are some possible combinations for wye and delta systems. Note that in each example, a neutral wire is always included.

FIGURE 12-1   How to label a three-wire system.

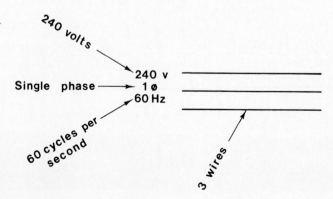

Three-Phase, Four-Wire Wye System

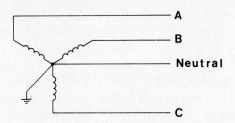

Used for apartment
buildings, office buildings,
shopping centers, etc.

Three-Phase, Three-Wire Delta System

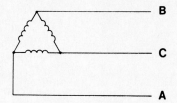

Used primarily for
factories or manufact-
uring shops.

FIGURE 12-2   Three-phase, four-wire wye system and a
three-phase, three-wire delta system.

TRANSFORMER   WINDINGS

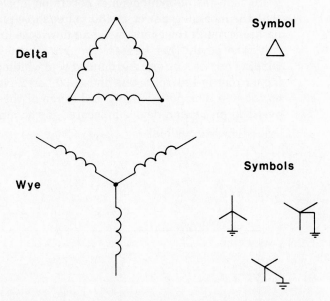

FIGURE 12-3   Transformer windings and their drawing
symbols.

FIGURE 12-4   How a single-phase, three-wire system
can be used to produce different voltages.

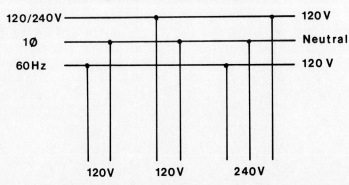

12-3   ONE-LINE DIAGRAMS

One-line diagrams are diagrams that graphically define the components
and the relationship between the components for an electrical circuit. The
components are arranged along a single vertical line which is read from
top to bottom. Actually, it is rare for a one-line diagram to consist of only
a single line; most often, several one-line diagrams are drawn together to
show the overall power distribution system, rather than just a single circuit. Figure 12-5 shows a drawing that consists of several one-line diagrams.

12-4   HOW TO READ
A ONE-LINE DIAGRAM

The basic symbols used to draw one-line diagrams are presented in Figure
12-6. It is suggested that these symbols be drawn and filed to create a
symbol library as defined in Section 4-2. Figure 12-7 shows a one-line diagram that has all symbols and notations labeled. We see in Figure 12-7
that the incoming power is 480 volts, three-phase, 60 hertz. The power
passes through the connection and down the line that represents the circuit. The power then passes through a transformer which has a delta
winding on one side and a grounded wye winding on the other. The resulting current is 120 volts, one phase, 60 hertz. Next, an ammeter is placed
on the line. The power rating and range of the ammeter are defined. An
overload protection device is located in the ammeter line to protect the
ammeter from damage.

FIGURE 12-5 Example of
three one-line diagrams which
are part of the same circuit.

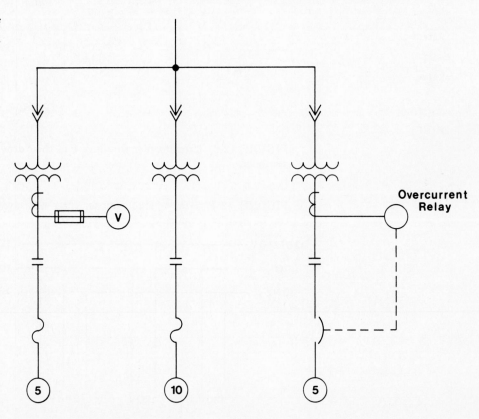

Overcurrent
Relay

SYMBOLS for ONE LINE DIAGRAMS

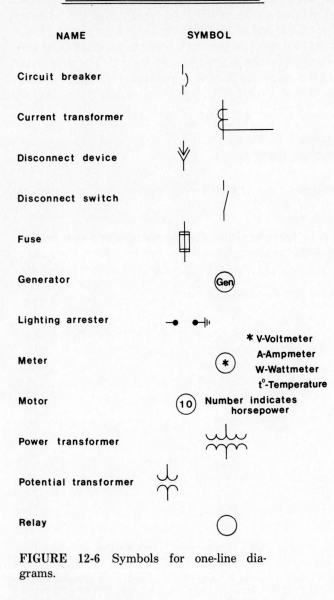

FIGURE 12-6 Symbols for one-line diagrams.

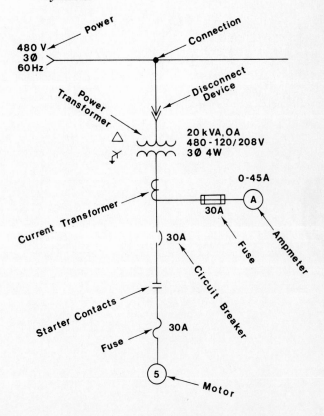

FIGURE 12-7 One-line diagram with labeled symbols.

The main circuit then continues through a circuit breaker, starter switch, another overload protection device, and finally into the 5-horsepower electrical motor at the end of the line.

It should be noted that the labeling on a one-line diagram is an important aspect of the diagram. The drafter should always carefully check *all* labeling used and then have the design engineer double-check the labeling.

12-5 HOW TO DRAW
A ONE-LINE DIAGRAM

To draw a one-line diagram, use the following procedure (illustrated in Figure 12-8):

1. Draw very light horizontal lines to represent the power source and a vertical line as shown.
2. Add all symbols for the various components.
3. Trim all excess lines and add labeling.

FIGURE 12-8 How to draw a one-line diagram.

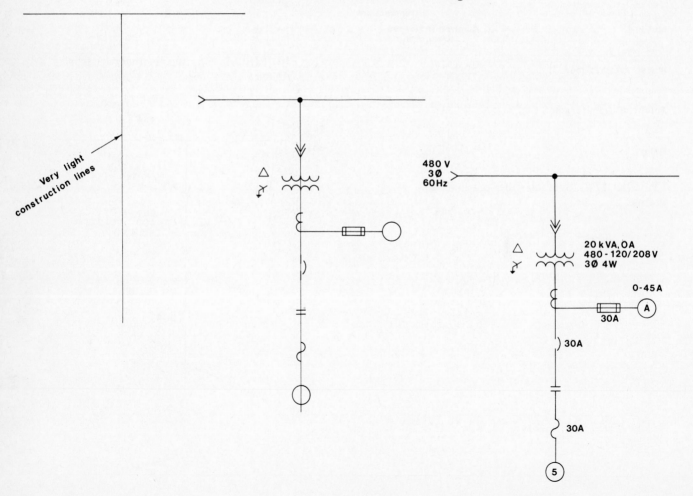

12-6   LADDER DIAGRAMS

Ladder diagrams are a type of diagram used to define industrial control circuits. They derive their name from the fact that they resemble ladders—two long parallel vertical lines connected by a series of horizontal runners. Figure 12-9 is an example of a ladder diagram.

Ladder diagrams are most commonly used to draw the circuitry required to activate and deactivate electrical motors and coils. These components, in turn, are used to operate elevators, heating and air-conditioning systems, machines, subway systems, and so on.

12-7   HOW TO READ A LADDER DIAGRAM

Ladder diagrams are drawn in two parts: the horizontal upper portion, which defines the power applications; and the vertical/horizontal ladder portion, which defines the control functions. The diagram is read from left to right across the power lines, then down the left side of the ladder portion and across each individual horizontal line from left to right. Figure 12-10 shows this concept.

FIGURE 12-9   Example of a ladder diagram.

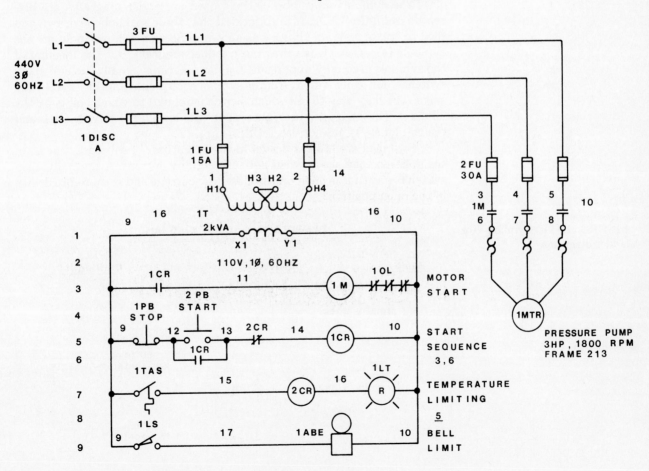

NOTES:
1. SET 1TAS FOR 180°F.

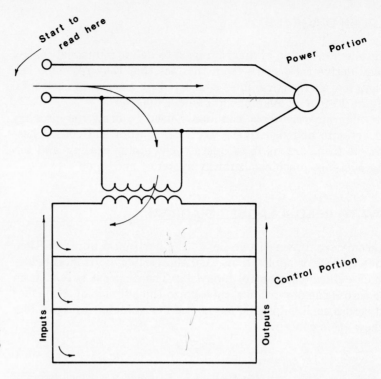

FIGURE 12-10 Ladder diagrams are read from left to right and top to bottom.

The symbols and abbreviations used on ladder diagrams are presented in Figure 12-11. It is suggested that these symbols be drawn and filed to create a symbol library as shown in Section 4-2. Not only are the symbols important when drawing a ladder diagram, but the labeling of the symbols is equally important. Ladder diagrams usually include many switches and coils which, from a symbol viewpoint, appear exactly the same. The only way to tell which switch is related to which coil is by the labeling. For example, in Figure 12-9 CR1 activates switch CR1 and the coil labeled CR2 activates CR2.

Switches are always drawn in their "natural" positions, that is, in the position they are in when not activated by current. A normally open switch is one that is open unless closed by current and is therefore drawn in the open position.

FIGURE 12-11 Symbols for ladder diagrams.

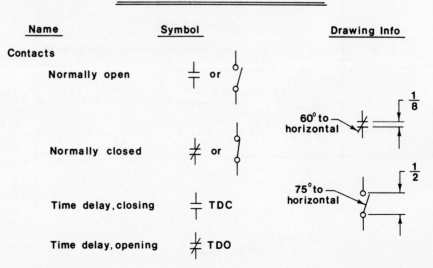

SYMBOLS for LADDER DIAGRAMS

Name	Symbol	Drawing Info
Contacts		
Normally open	$\dashv\vdash$ or	
Normally closed	$\dashv\!\!\!/\vdash$ or	60° to horizontal 1/8
Time delay, closing	$\dashv\vdash$ TDC	75° to horizontal 1/2
Time delay, opening	$\dashv\!\!\!/\vdash$ TDO	

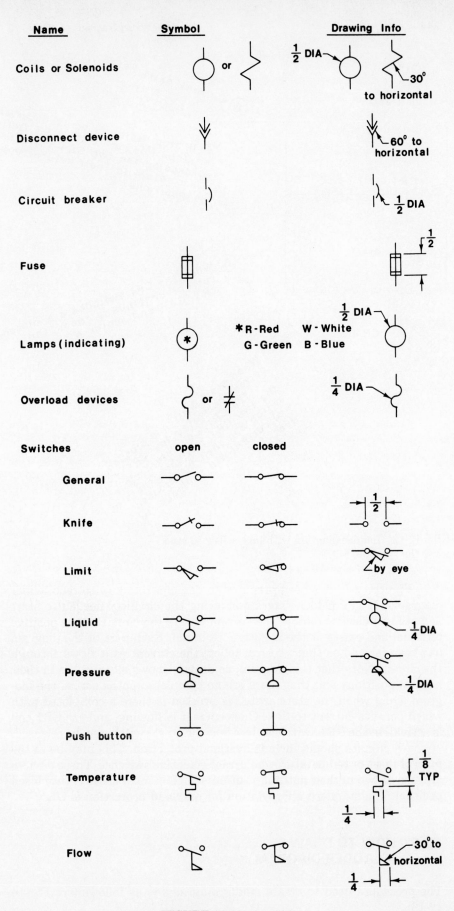

Name	Symbol		Drawing Info

Coils or Solenoids — or — ½ DIA — 30° to horizontal

Disconnect device — 60° to horizontal

Circuit breaker — ½ DIA

Fuse — ½

Lamps (indicating) — *R - Red W - White G - Green B - Blue — ½ DIA

Overload devices — or — ¼ DIA

Switches open / closed

General

Knife — ½

Limit — by eye

Liquid — ¼ DIA

Pressure — ¼ DIA

Push button

Temperature — ⅛ TYP — ¼

Flow — 30° to horizontal — ¼

FIGURE 12-11 *(cont'd)*

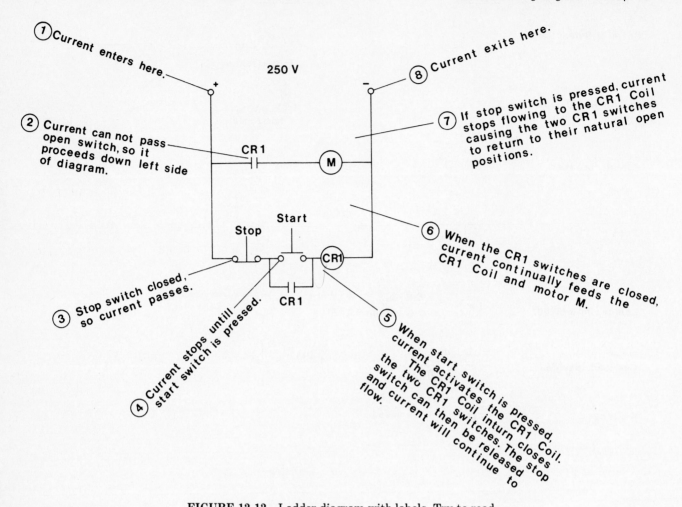

FIGURE 12-12   Ladder diagram with labels. Try to read the story the symbols tell.

Figure 12-12 has been prepared to demonstrate how to read ladder diagrams. It is a ladder diagram showing the circuitry needed to start and stop an electric motor. The current enters the circuit at the positive terminal and exists at the negative terminal. Each part of the diagram has been labeled so that you can follow the current as it flows through the circuit. Note that the diagram, as drawn, shows all switches in their natural positions and that there are no completed paths across the diagram. Only when the start switch is pressed is there a completed path which permits current to flow. Once current is flowing, and the CR1 coil is activated, the CR1 switches close and the start switch may be released.

All circuits should include overload protection. This may be in the form of fuses or temperature- or current-sensitive switches. These devices are also drawn in their natural position, which in most cases is the closed position. The standard abbreviation for overload protection is OL.

12-8   HOW TO DRAW A LADDER DIAGRAM

The procedure used to create ladder diagrams is as follows (see Figure 12-13):

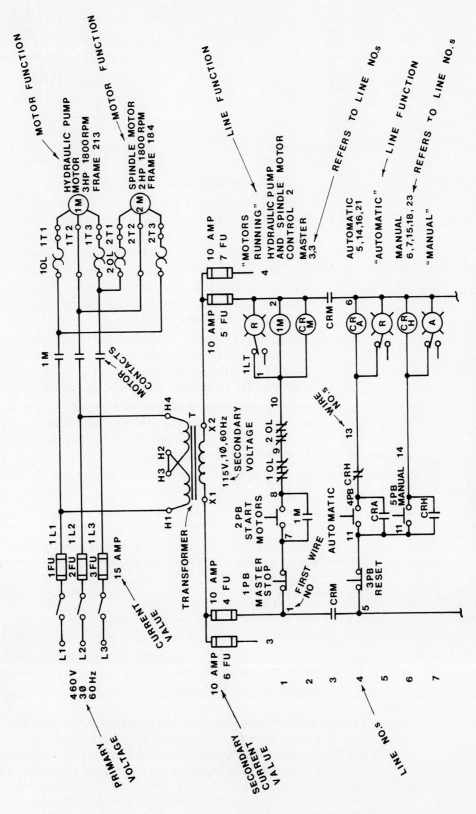

FIGURE 12-13   How to draw a ladder diagram.

245

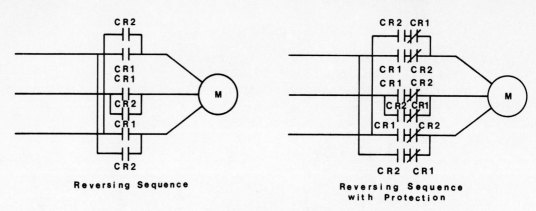

FIGURE 12-14   Reversing sequence.

1. Draw in the horizontal power lines. If two motors are required, use the arrangement shown in Figure 12-13. If a reversing sequence is required, use the setup shown in Figure 12-14.

2. Add the disconnect switches, fuses, and starter contacts as shown. The overload devices shown near the motor are usually manufactured as part of the motor. Nevertheless, they must be included on the diagram.

3. Add the transformer. Transformers that convert 440 volts to 110 volts have terminals H3 and H2 connected as shown in Figure 12-13. Transformers that convert 440 volts to 220 volts connect pins H1 to H3 and H2 to H4, as shown in Figure 12-15.

4. Lay out the control portion of the diagram. There is no standard spacing, but placing the vertical lines 6 inches apart and the horizontal lines 1 inch apart will usually be sufficient. Larger diagrams require larger spacing.

5. Add the required symbols, trim any excess lines, and label the diagram.

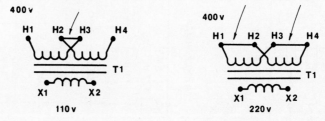

FIGURE   12-15 Transformer
440 volts to 220 volts.

12-9 HOW TO LABEL A LADDER DIAGRAM

Figure 12-13 shows a labeled ladder diagram. Each label has been identified by a note. Specific labeling will vary from company to company depending on the product being made and individual preferences.

The diagram used is part of an example presented by the Joint International Council (JIC) of 7901 Westpark Drive, McLean, Va. 22102 in their publication *Electrical Standards for Mass Production Equipment* EMP-1-67. The JIC standards are generally accepted and used throughout the electrical industry.

12-10   RISER DIAGRAMS

Riser diagrams are drawings that show the wiring paths of a building's electrical system up to, but not including, the branch circuits. Figure 12-16 is an example of a riser diagram which shows the between-floor wiring paths of a security system. The size of the blocks may be varied according to individual needs, but the distance between floor lines is generally kept equal.

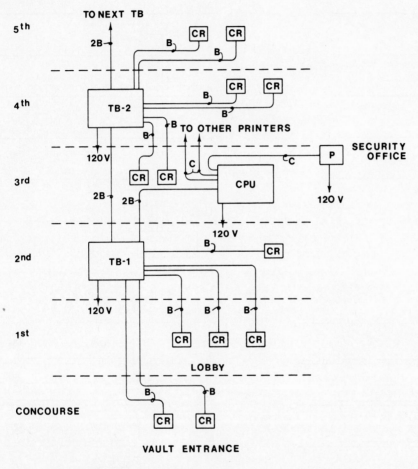

FIGURE 12-16   Example of a riser diagram.

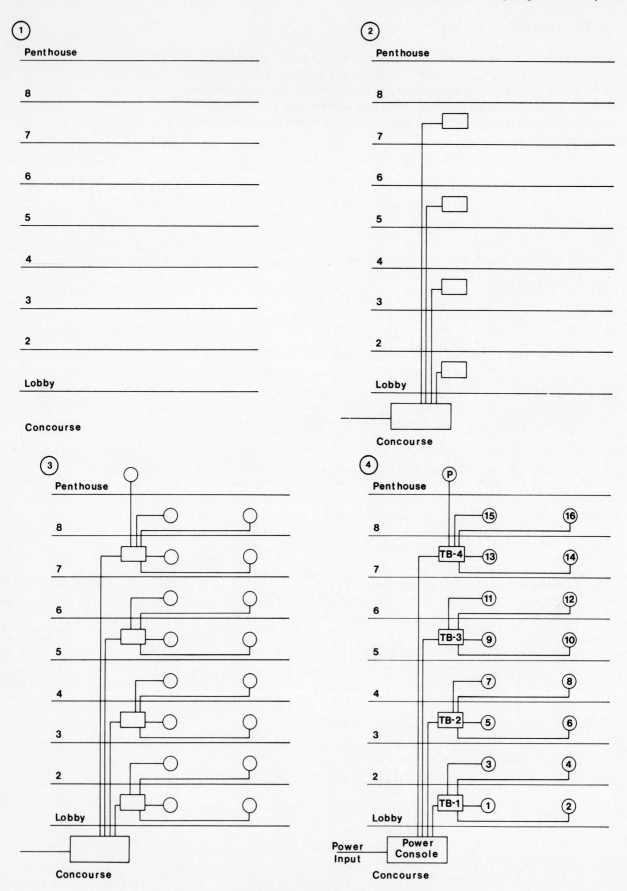

FIGURE 12-17 How to create a riser diagram.

12-11 HOW TO READ
A RISER DIAGRAM

Riser diagrams are read by starting at the power input point and then following each wiring path to its destination. In Figure 12-16 we see that the power enters the building in the basement. It then proceeds to the first security center console and from there to the individual stations.

The legend in the lower right-hand corner of the drawing defines all the symbols used. All symbols should be clearly defined even if their meaning seems to be obvious.

Each floor must be labeled and separated by a horizontal line. Special or large important components such as the security center console and the main power console are labeled on the drawing as shown. They should not be abbreviated.

12-12 HOW TO DRAW
A RISER DIAGRAM

Figure 12-17 illustrates how to create a riser diagram. The diagram was originally presented in sketch form in Figure 12-18. The procedure used is as follows:

1. Study the information and requirements to determine how many floors are to be drawn and what components are needed. Draw the basic floor pattern so that floors are evenly spaced. If a floor has a large number of components, its size may be increased.

2. Draw in all major components.

3. Draw in all minor components and add wiring paths.

4. Label all components and set up a legend. Every component used must be defined in the legend.

5. Trim all excess lines.

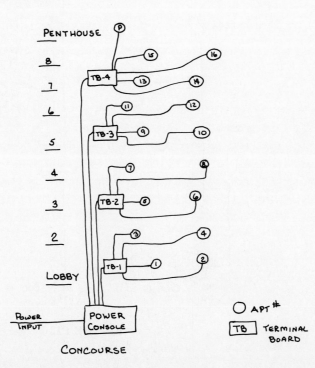

FIGURE   12-18 Freehand sketch of a riser diagram from which Figure 12-15 was prepared.

Problems

P12-1   Prepare a one-line diagram of the sketch shown in Figure P12-1.

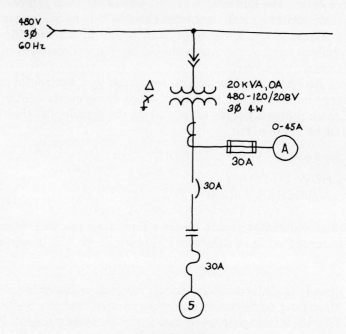

FIGURE P12-1

P12-2   Prepare a one-line diagram of the sketch shown in Figure P12-2.

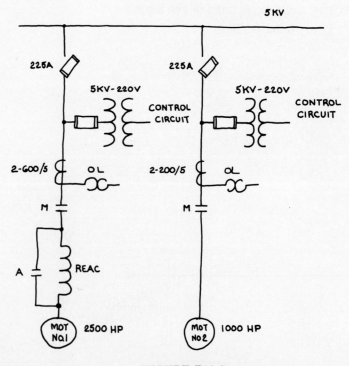

FIGURE P12-2

P12-3   Prepare a one-line diagram of the sketch shown in Figure P12-3.

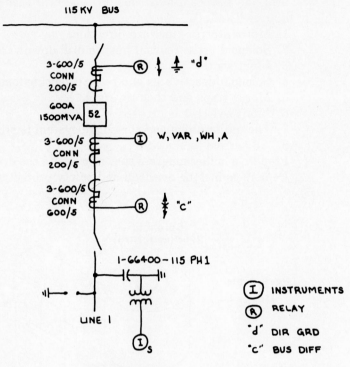

FIGURE P12-3

P12-4   Redraw the ladder diagram shown in Figure P12-4 and substitute the following symbols:

1. Fuse L1
2. Fuse L2
3. Contact, normally open, CR1
4. A motor M
5. An overload device, OL
6. Transformer, 115 V to 60 V
7. An overload device, OL
8. A start/stop sequence which includes a start button, a stop button, a coil labeled CR1, and a normally open switch labeled CR1
9. An overload device, OL

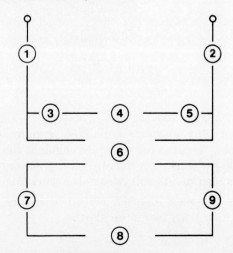

FIGURE P12-4

P12-5   Figure P12-5 shows a drill press that has been rigged for auto-
        matic operation. The operator positions the workpiece under the
        drill and pushes a start button. The drill then follows the se-
        quence listed.
        1. Motor starts — forward direction.
        2. Solenoid activates and pushes drill downward.
        3. Limit switch 2 is pressed.
        4. Solenoid deactivates and returns (spring-loaded return).
        5. Motor stops and then reverses.
        6. Limit switch 1 is pressed.
        7. The motor stops and the entire system is stopped.
        8. All current is shut off.
        Prepare a ladder diagram that represents the drilling sequence.
        Be sure to include overload and safety protection devices.

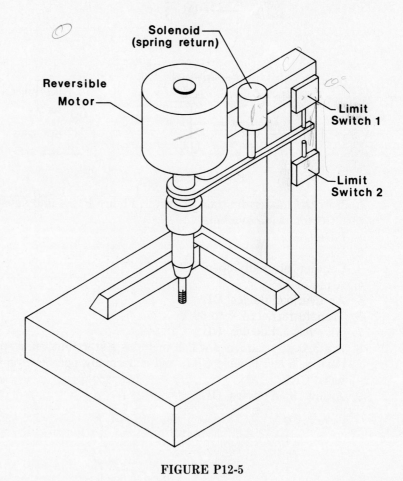

FIGURE P12-5

P12-6   This problem is exactly like Problem P12-5 except that in this
        case two more solenoids are added as clamping devices. These
        solenoids will press the workpiece against the guide rails shown
        and hold the workpiece during drilling. After the drilling is com-
        plete, the solenoids will release the workpiece.

P12-7   Redraw the riser diagram shown in Figure P12-7 and add a
        fourth floor with two more alarms.

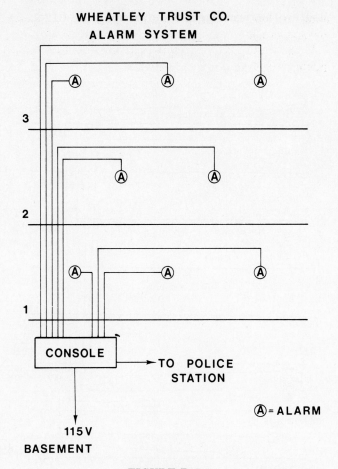

WHEATLEY TRUST CO.
ALARM SYSTEM

FIGURE P12-7

P12-8   Redraw the riser diagram shown in Figure P12-8.

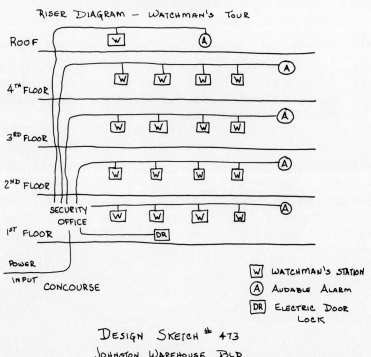

FIGURE P12-8

P12-9   Redraw the riser diagram shown in Figure P12-9.

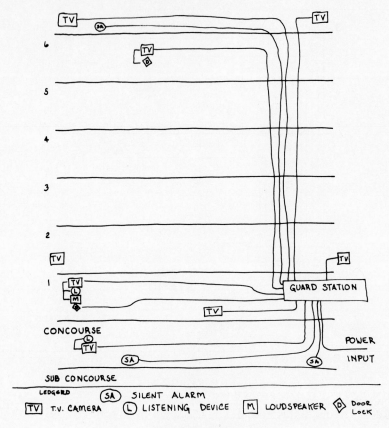

FIGURE P12-9

P12-10   Redraw Figure 12-8 approximately twice as large as presented.

P12-11   Redraw Figure 12-12 approximately twice as large as presented.

P12-12   Design a circuit that will raise and lower the dumbwaiter as shown in Figure P12-12. The dumbwaiter must be able to be called from either station.

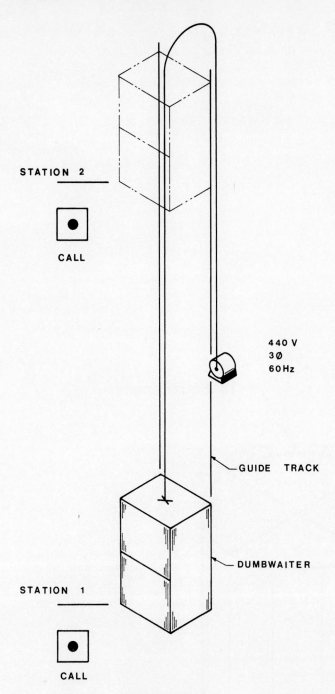

STATION 2

CALL

440 V
3∅
60 Hz

GUIDE   TRACK

DUMBWAITER

STATION 1

CALL

FIGURE P12-12

P12-13   Repeat Problem P12-12 but include automatic door openers and a safety system that prevents a person's hand from being jammed in the door.

13

Graphs *and Charts*

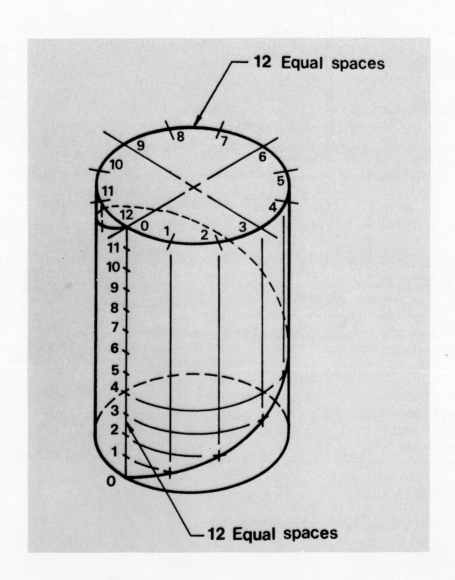

13-1 INTRODUCTION

This chapter explains how to create graphs and charts. Drafters are often asked to convert lists of raw data into graph or chart form so that the data may be more easily understood. Most engineering test data, for example, originally appear in computer printout form — long, long lists of numbers. To read and analyze these lists is tedious work. If, however, the same data are presented in graph form, it is much easier to understand.

This chapter covers pie charts, bar charts, and curve plotting. Logarithmic and semilogarithmic grids and scales are also included.

13-2 PIE CHARTS

Pie charts are used to illustrate the relative sizes of various component parts of a total quantity. The name pie chart comes from its shape, which looks like an overview of a pie which has been sliced into different-sized pieces. Figure 13-1 illustrates a pie chart.

To demonstrate how to create pie charts, consider the following problem (Figure 13-2 illustrates). You are asked to draw a pie chart that will illustrate the relative numbers of persons serving the Army, Navy, Air Force, and Marine Corps in 1975. You are given the following figures:

Military Personnel, 1975

Army	781,316
Navy	549,400
Air Force	608,337
Marines	192,200

Production of Electricity in U.S. by Source — 1975

FIGURE 13-1 Example of a pie chart.

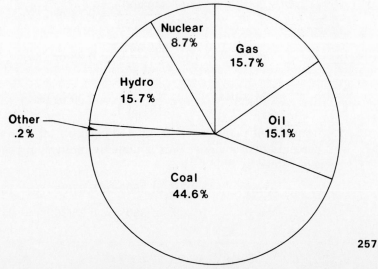

**Persons Serving in the
Army, Navy, Air Force, & Marines
1975**

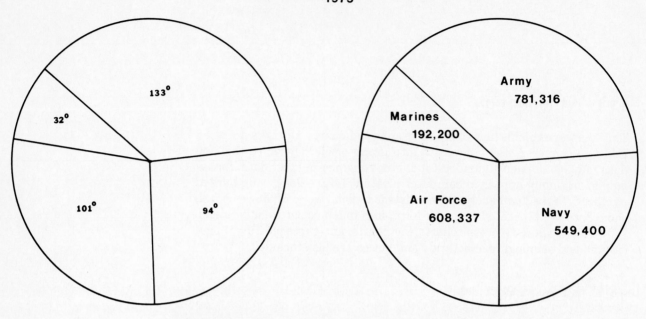

FIGURE 13-2   How to prepare a pie chart. The degree
values are based on calculations.

To convert these figures into a pie chart:

1. Calculate the percentage value of each figure. This is done by
    first finding the total value of all the figures and then dividing
    each figure into the total.

$$
\begin{array}{l}
781{,}316 \\
549{,}400 \\
608{,}337 \\
\underline{192{,}200} \\
2{,}131{,}253
\end{array}
\qquad \text{Total value}
$$

Army $\dfrac{781{,}316}{2{,}131{,}253} = 0.37$ or 37%

Navy $\dfrac{549{,}400}{2{,}131{,}253} = 0.26$ or 26%

Air Force $\dfrac{608{,}337}{2{,}131{,}253} = 0.28$ or 28%

Marines $\dfrac{192{,}200}{2{,}131{,}253} = 0.09$ or 9%

2. Convert the percentages calculated in step 1 into the equivalent
    percentage of a circle. This is done by multiplying the percent-
    age values by 360°:

Army   37% of 360° = 0.37(360°) = 133°

Navy   26% of 360° = 0.26(360°) = 94°

Air Force   28% of 360° = 0.28(360°) = 101°

Marines   9% of 360° = 0.09(360°) = 32°

These values are in terms of degrees, and can be measured using a protractor.

3. Lay out a circle about 6 inches in diameter and mark off the degree values calculated in step 2.

4. Label each sector as shown and add a title for the chart. If desired, the different sectors may be shaded differently to help distinguish them from each other.

Many software packages are available to draw both two-dimensional and three-dimensional pie charts. They are often found in software for business graphics.

13-3 BAR CHARTS

Bar charts are used to demonstrate the differences between fixed quantities. They derive their name from the fact that they express values in terms of bar-shaped figures, as shown in Figure 13-3.

It is important that bar charts be drawn with a properly proportioned scale. Scales that are too small are difficult to read and may not show the differences between values clearly, while scales that are too large may not fit the paper.

FIGURE 13-3 Example of a bar chart.

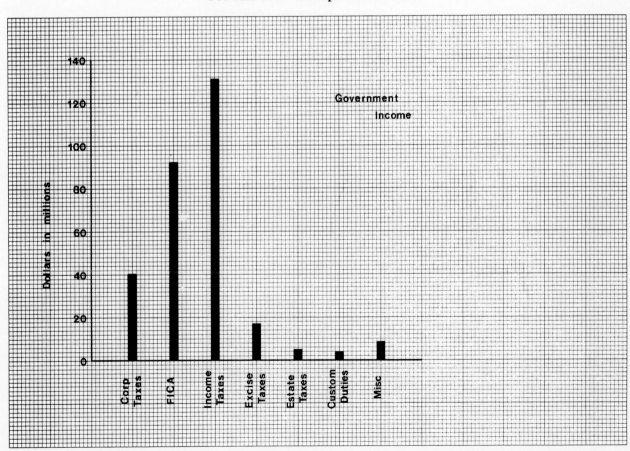

Because bar charts usually involve a comparison of fixed values, the values are sometimes lettered in just above the top of the bar. In this way the viewer gets not only a visual comparison but also the numerical values so that further comparison may be made if desired. This information could be derived from the bar chart scale, but it is usually very helpful to the reader to have the values clearly stated.

Consider the following problem (Figure 13-4 illustrates). You are asked to create a bar chart that compares the number of home runs hit by the National League's leading home-run hitters from 1920 to 1924. The number of home runs hit was:

Year	Player	Home runs
1920	Cy Williams, Philadelphia	15
1921	George Kelly, New York	23
1922	Rogers Hornsby, St. Louis	42
1923	Cy Williams, Philadelphia	41
1924	Jacques Fourmier, Brooklyn	27

To convert these figures into a bar chart:

1. Choose a scale. In this case we let each block on the vertical axis represent 1 home run, or 10 home runs to the inch. Whenever possible, choose a scale that is directly related to 10, as most people are used to thinking in terms of a base 10.

2. Define the horizontal and vertical axis.

3. Draw the bars.

4. Label each bar and add the title and any other necessary information.

FIGURE 13-4   How to prepare a bar chart.

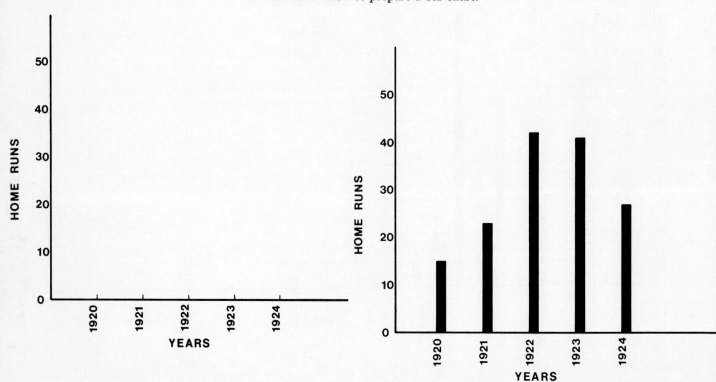

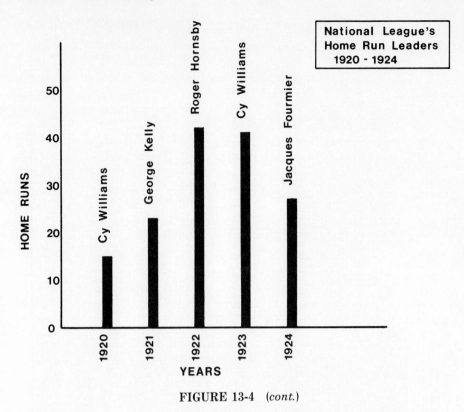

FIGURE 13-4   *(cont.)*

It is not necessary to have the horizontal axis represent a value of 0. If there are extremely large values to work with, the length of the bars may be reduced by assigning a value higher than 0 to the horizontal axis.

13-4   CURVE PLOTTING

Curve plotting is a term used to describe the creation of graphs that are based on data that are a function of two or more variables (Figure 13-5 illustrates). Curve plotting does not produce a comparison between fixed values as do pie charts and bar graphs, but instead shows the relationship between two or more different variables.

For example, in Figure 13-5, we see that the horizontal axis is a measure of engine speed (RPM) and the vertical axis is a measure of horsepower (HP). Both engine speed and horsepower are variables. The shape of the curve shows the relationship between the two variables. The curve slopes upward, meaning that as values of engine speed increase, values of horsepower also increase, up to the point of maximum horsepower. After this point, as the values of engine speed increase, the values of horsepower decrease — the curve slopes downward. This means that after we pass the point of maximum horsepower, increasing the engine speed actually reduces the amount of horsepower the engine generates.

Four areas often cause trouble when plotting curves: choosing a scale, identifying the curves properly, drawing a smooth curve between the data points, and choosing the proper grid background (sometimes patterns other than squares are used). Each area will be considered separately.

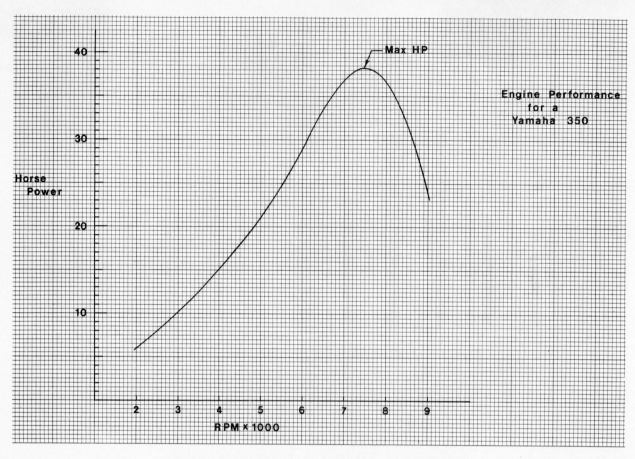

FIGURE 13-5   Example of a curve that shows the relationship between two or more variables, in this case horsepower versus rpm.

13-5   CHOOSING A SCALE

A scale must be chosen so that it will be easy to understand. It is tempting to pick a scale which is mathematically convenient — that is, the scale is easy to fit to the data, but such scales are usually very difficult to read and work with. For example, consider the two curves plotted in Figure 13-6. Both represent the same data, but the curve on the left uses an odd scale and the curve on the right uses a scale based on 10. What are the x and y values of point A? If we use the curve on the right we can easily find the answer $x = 8.0$, $y = 13.3$, but we see how much more difficult it is to derive the same answer from the left curve.

A scale must also consider the size of the paper on which the curves are to be drawn. Most 8½ by 11 inch sheets of graph paper have a grid pattern which is 8 by 10 inches (some types of graph paper have even smaller grids). The workable area of the paper is further reduced when the space needed to define the horizontal and vertical axes is taken into account. This means that the final workable area — the area in which we can draw curves — for an 8½ by 11 inch sheet of paper is only about 7 by 9 inches at most (see Figure 13-7). To choose a scale which is both easy to read and will fit the paper, consider the following problem.

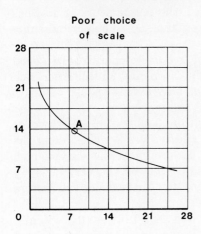

Poor choice
of scale

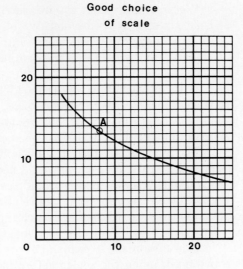

Good choice
of scale

FIGURE 13-6 Care must be taken when choosing a scale. What is the value of point A?

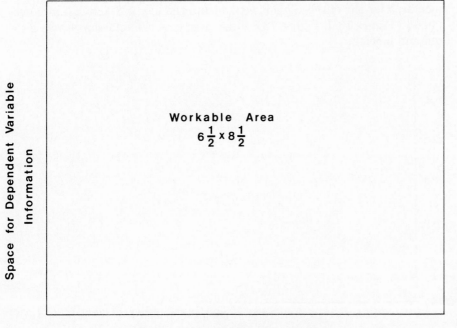

Space for Dependent Variable Information

Workable Area
$6\frac{1}{2} \times 8\frac{1}{2}$

Space for Independent Variable Information

FIGURE 13-7 Always leave room for identifying the axis. This means that the area for curve plotting is actually less than the total grid pattern.

Plot the performance curve for a Texas Instruments SN52506 dual differential comparator with strobe as a function of input offset current versus the free air temperature. The test data are as follows:

Free air temperature (°C)	Input offset current (μA)
−50	2.09
−25	1.40
0	0.94
25	0.67
50	0.53
75	0.43
100	0.36

We first notice that the free air values are equally spaced and that at least six spaces are needed. We also realize that the free air temperature values are the independent variables (the temperature of air does not depend on the values of the comparator) and should be plotted along the horizontal axis.

If we let each inch represent 25°C of free air temperature, we need at least 6 inches along the horizontal axis. Normally, 25 units per inch would not be a good choice, but because we have evenly spaced values it is acceptable in this situation.

The input offset current values vary from 0.36 to 2.09 μA. If we let each inch equal 1.00 μA, we would need a little over 2 inches along the vertical axis. However, this would result in an almost flat curve which would not clearly represent the relationship between the variables. By doubling the scale so that two blocks equals 0.1 μA, the curve will require a little over 4 inches on the vertical scale.

It is customary to extend curves beyond their furthest data points, so we need, using the scales determined above, approximately a 5 × 7 inch area for the curve. This is well within the 6½ × 8½ inch limits outlined in Figure 13-7. Figure 13-8 illustrates how the final curve was plotted and labeled.

FIGURE 13-8 How to prepare a curve plot: (a) define axis and label; (b) add data points and draw in curve;

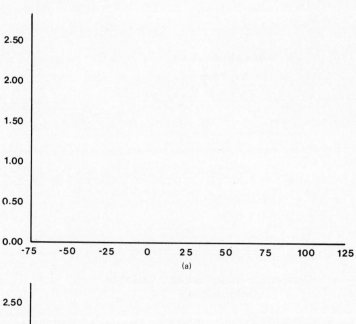

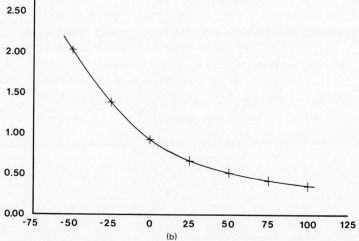

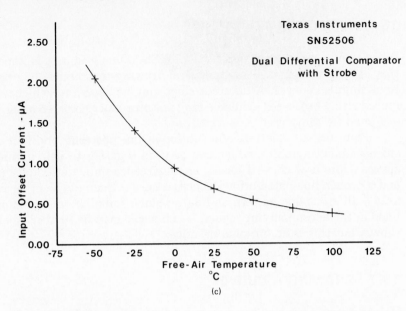

FIGURE 13-8(c)   Add all labels required.

Figure 13-9 plots age versus height for males and females. Because the curves are almost identical up until age 12, a large vertical scale is required to bring out the small differences. Note how small lines were added to both the horizontal and vertical axes to make it easier to understand the scale used.

Note how in all curve plots the final graph includes completely defined horizontal and vertical scales, including labels, a title, a definition of data points and line patterns, any notes required, and labels for all curves.

FIGURE 13-9   Sample curve plot of age versus height for males and females.

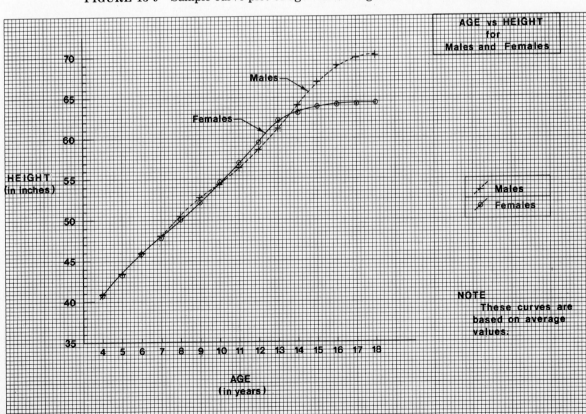

13-6   IDENTIFYING CURVES

When two or more curves are drawn on the same graph, it is important that each be clearly identified. Several drawing techniques may be used to distinguish curves. A different color may be used for each curve, but reproduction processes available may not reproduce color, so other methods must be considered.

Figure 13-10 illustrates four different line patterns, and four different data point symbols, which can be used together to help distinguish curves. Note how curve 1 uses a small circle to indicate its data points and a broken-line pattern for the actual curve. Each of the other curves uses a different line pattern and data point symbol, all of which are defined in the accompanying legend. Each curve may be further labeled as shown, but this is an optional practice.

13-7   DRAWING A CURVE

Curves drawn by a CAD system are called *splines*. Splines can be either open or closed, as shown in Figure 13-11.

SPLINE functions operate by first defining a series of points, usually by inputting coordinate values. The points are then joined by a smooth curve as shown in Figure 13-12.

Figure 13-13 shows two different types of average curves. The straight-line average is created by drawing a straight line through the data points so that approximately half the points are above the line and half below the line. The distance between the points above the line should be approximately equal to the distance between the line and the points below the line. Curved-line approximations can be made in a similar manner.

FIGURE 13-10   When plotting different curves on the same axis, use different line patterns and different data point symbols to help the reader distinguish the curves.

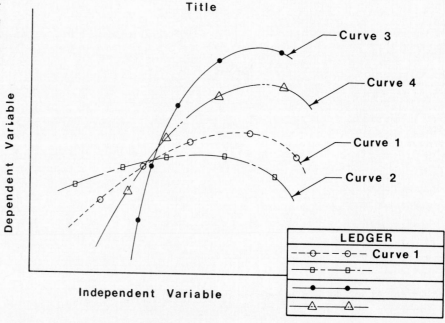

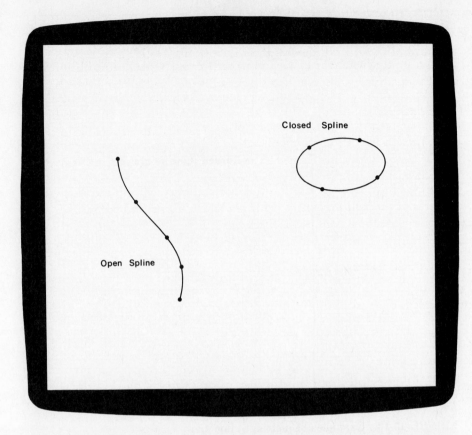

FIGURE 13-11 Examples of open and closed splines.

FIGURE 13-12 An example of a spline through known points.

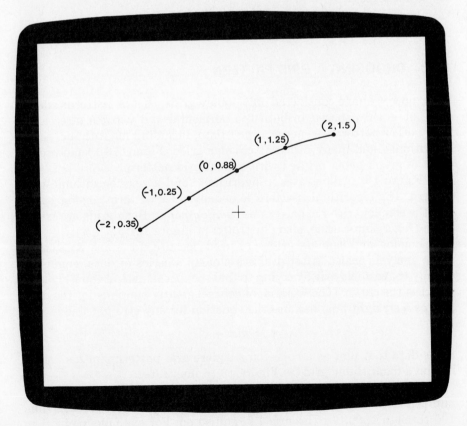

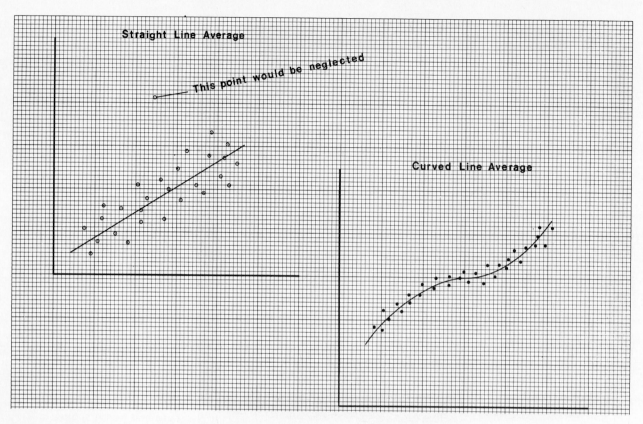

FIGURE 13-13 Examples of a straight-line and a curved-line average.

13-8 CHOOSING A GRID PATTERN

Up to now, we have considered only square grids, that is, patterns whose lines are evenly spaced in both the horizontal and vertical directions. Many other patterns are used to plot data, including logarithmic, semilogarithmic, and many different circular grids. Figure 13-14 pictures a sheet of graph paper printed with a polar grid pattern.

Figure 13-15 illustrates a logarithmic and a semilogarithmic grid pattern. The logarithmic pattern is usually referred to as *log-log paper* and the semilogarithmic pattern as *semilog paper*. Both grids are based on the logarithmic scale, also illustrated in Figure 13-15.

Log-log and semilog grids are used because, for some data, they make analysis easier. After drafters plot up a series of data, engineers usually try to analyze it by trying to find the mathematical equation that matches the curve. The work is, of course, greatly simplified if the data plot as a straight line, because the equation for any straight line is

$$y = mx + c$$

Some data that plot as a curve on a square grid pattern plot a straight line on a logarithmic grid, as Figure 13-16 illustrates.

The basic logarithmic grid can represent any multiple of 10 of the numbers 1 through 10. It can represent 1 through 10, 10 through 100, 100 through 1000, 0.1 through 1.0, and so on. For example, point *A* in

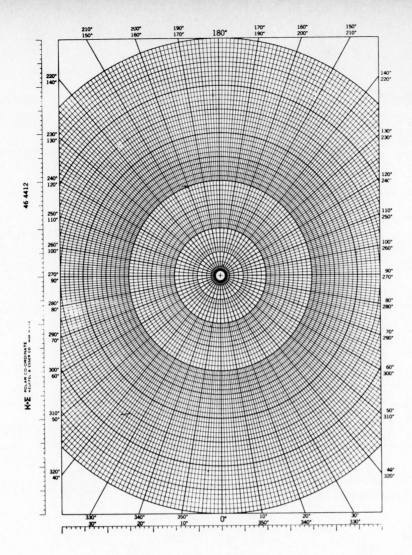

FIGURE 13-14 Sample sheet of polar graph paper.

FIGURE 13-15 Samples of a logarithmic and semilogarithmic grid pattern and a sample logarithmic scale.

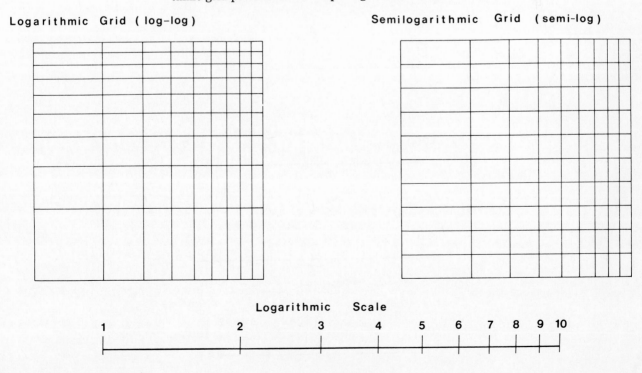

Logarithmic Grid (log-log)

Semilogarithmic Grid (semi-log)

Logarithmic Scale

1   2   3   4   5   6 7 8 9 10

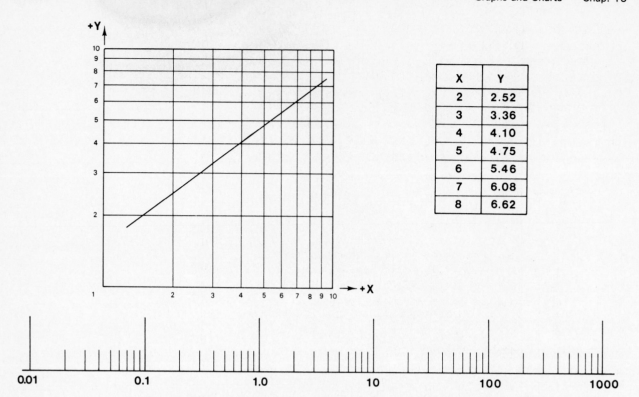

X	Y
2	2.52
3	3.36
4	4.10
5	4.75
6	5.46
7	6.08
8	6.62

FIGURE 13-16 Sample of data points that plot as a straight line on a log-log pattern.

FIGURE 13-17 How to read different log scales.

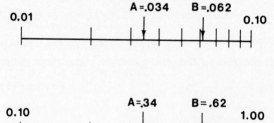

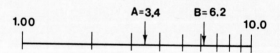

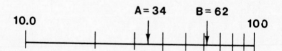

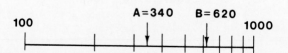

Figure 13-17 can equal 0.034, 0.34, 3.40, 34, or 340 depending on which scale is used. Similarly, point B can equal 0.062, 0.62, 6.20, 62, or 620.

A unique feature of log grids is that they can *never* have a line with a value equal to 0.0. A line may be defined as 0.00001 or even smaller, but it can never be 0.0. This means that curves drawn on log grids can never cross, or even touch, the vertical axis.

Log scales may be placed one after another as shown in Figure 13-17 to increase the overall value range of the graph. It is good practice to label at least the beginning and end values of each log pattern used.

Problems

P13-1   Figure P13-1 illustrates three different mathematical curves, that is, curves whose data points were obtained by substituting values into mathematical equations. In each example the x variables are considered the independent variable and the y variables are considered the dependent variable. Plot the following equations as assigned by your instructor. Include your calculation sheets. Always define the x variable as the independent variable.

(a) $y = x$

(b) $y = 2x$

(c) $y = -\frac{1}{2}x$

(d) $y = 1.6x + 2$

(e) $4y = 2x - 1.50$

(f) $y = x^2$

(g) $y = x^2 + 2x$

(h) $y = \frac{1}{2}x^2 + 2x + 4$

(i) $y = 1.6x^2 - 2.2x - 0.7$

(j) $2y = -x^2 + 3x - 0.5$

(k) $xy = 4$

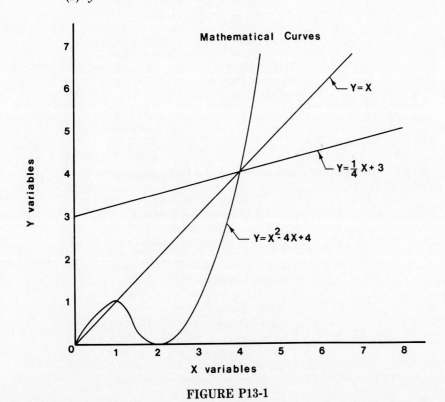

FIGURE P13-1

P13-2   Draw a pie chart based on the following data:

World Power Consumption, 1976

Country	Kilowatt-hours
United States	1,999,688
USSR	1,038,000
Japan	432,757
West Germany	301,800
Others	2,480,755
World total	6,253,000

P13-3   Draw a pie chart based on the following data:

U.S. Government Crime Report, 1977

Murder	20,510
Forcible rape	56,090
Robbery	464,970
Aggravated assault	484,710
Burglary	3,252,100
Larceny-theft	5,977,700
Auto theft	1,000,000
Total	11,256,000

P13-4   Draw a bar chart based on the following data:

Electricity Generated
by Atomic Power

Country	Output (MW)
United States	28,236
Great Britain	4,258
USSR	4,521
Japan	3,718
France	2,886

P13-5   Draw a bar chart based on the following data:

Leading Lifetime Scores
in Professional Football
through 1977

Name	Points
George Blanda	2002
Lou Groza	1608
Fred Cox	1227
Jim Bakken	1171
Jim Turner	1147
Gino Cappelletti	1130
Bruce Gossett	1031

P13-6   Draw a bar chart based on the following data:

Leading Rebounders in NBA, 1975–1976

Name	Team	Rebounds
Abdul-Jabbar	Los Angeles	1383
Cowens	Boston	1246
Unseld	Washington	1036
Silas	Boston	1025
Lacey	Kansas City	1024

P13-7   Plot the following data. Plot the years along the horizontal axis and the hourly earnings along the vertical axis.

Increases in Factory Workers'
Hourly Earnings

Year	Hourly earnings
1965	2.61
1966	2.72
1967	2.83
1968	3.01
1969	3.19
1970	3.36
1971	3.56
1972	3.81

P13-8   Ohm's law defines the relationships among current, voltage, and resistance as

$$I = \frac{E}{R}$$

where I = current (amperes)

E = voltage (volts)

R = resistance (ohms)

Plot a family of curves comparing amperes and volts at four different values of ohms. Plot the volts along the horizontal axis.

The values to be plotted may be calculated by first assuming a constant value for ohms and then inserting various values for volts and solving the equation for amperes. For example, assume that the ohms value equals 2 ohms. If we solve the equations for voltage values of 1, 2, 3, 4, and 5 we have

$$I = \frac{1}{2} = 0.5$$

$$I = \frac{2}{2} = 1.0$$

$$I = \frac{3}{2} = 1.5$$

$$I = \frac{4}{2} = 2.0$$

$$I = \frac{5}{2} = 2.5$$

We can then plot the values:

Volts	Amperes
1	0.5
2	1.0
3	1.5
4	2.0
5	2.5

Label each curve as to its constant ohms value.

P13-9   Plot the following equations on semilog paper.

(a) $y = x$ (e) $y = e^x$

(b) $y = 3x + 2$ (f) $y = e^{x+2}$

(c) $y = x^2$ (g) $y = e^x + 0.60$

(d) $y = x^2 + 1.5x - 4$

P13-10   Plot the following equations on log-log paper.

(a) $y = x$ (d) $y = e^{x+2}$

(b) $y = -x^2 + 2.4$ (e) $y = e^x + 2$

(c) $y = e^x$

Appendices

A

Digital Readout Letters

DIGITAL READOUTS

Display
Shape

Basic Letter Shapes

0 1 2 3 4

5 6 7 8 9

Variations

8 3 4

8 3 4

276

B

Resistor Color Codes

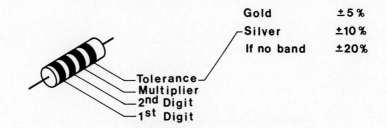

	Gold	±5%
	Silver	±10%
	If no band	±20%

Tolerance
Multiplier
2nd Digit
1st Digit

Color	1st Digit	2nd Digit	Multiplier
Black	0	0	1
Brown	1	1	10
Red	2	2	100
Orange	3	3	1,000
Yellow	4	4	10,000
Green	5	5	100,000
Blue	6	6	1,000,000
Violet	7	7	10,000,000
Gray	8	8	100,000,000
White	9	9	1,000,000,000
Gold	–	–	.1
Silver	–	–	.01

— Gold
— Orange
— Blue
— Red

2600 ohms ± 5%

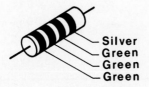

— Silver
— Green
— Green
— Green

550,000 ohms ±10%

277

C

Wire and Sheet Metal Gages

WIRE AND SHEET METAL GAGES			
Gage	Thickness	Gage	Thickness
000 000	0.5800	18	0.0403
00 000	.5165	19	.0359
0 000	.4600	20	.0320
000	.4096	21	.0285
00	.3648	22	.0253
0	.3249	23	.0226
1	.2893	24	.0201
2	.2576	25	.0179
3	.2294	26	.0159
4	.2043	27	.0142
5	.1819	28	.0126
6	.1620	29	.0113
7	.1443	30	.0100
8	.1285	31	.0089
9	.1144	32	.0080
10	.1019	33	.0071
11	.0907	34	.0063
12	.0808	35	.0056
13	.0720	36	.0050
14	.0641	37	.0045
15	.0571	38	.0040
16	.0508	39	.0035
17	.0453	40	.0031

D
Paper Sizes

PAPER SIZES

Size	Dimensions
A	$8\frac{1}{2}$ x 11
B	11 x 17
C	17 x 22
D	22 x 34
E	34 x 44
J	Roll Size

E
Wire Color Codes

COLOR CODES		
COLOR	LETTER CODE	NUMBER CODE
Black	BK	0
Brown	BR	1
Red	R	2
Orange	O	3
Yellow	Y	4
Green	GR	5
Blue	BL	6
Violet	V	7
Gray	GY	8
White	W	9

F

Standard Thread Sizes

Whenever possible, drafters should call for standard thread sizes in their designs. Standard threads may be purchased from many different manufacturers, are completely interchangeable and are relatively inexpensive when compared to "special" thread sizes.

Tables F-1 and F-2 are the UNC and UNF standards. To find the standard size for a given diameter, look up the diameter under the desired thread (UNC or UNF) and read the standard thread size adjacent to it. For example, ¼-diameter thread UNC has 20 threads per inch. The drawing callout would be

$$\frac{1}{4}\text{-20 UNC}$$

A 1¼ UNF has 12 threads per inch and would be called out on a drawing as

$$1\frac{1}{4}\text{-12 UNF}$$

The size numbers at the top of the tables are for small diameter threads. For example, a No. 4 UNF has a diameter of 0.112 and 48 threads per inch. The drawing callout would be

$$\#4(0.112)\text{-48 UNF}$$

Tables F-3 and F-4 define the 8 and 12 National (N) series thread. In each case, all diameters in the series are made with the same number of threads. All 8 series threads have 8 threads per inch. All 12 series threads have 12 threads per inch. For example, a 1⅞ diameter series 8 thread would have the drawing callout

$$1\frac{7}{8}\text{-8 UN}$$

TABLE F-1
UNC

Diameter	Threads per inch, P
1 (0.073)	64
2 (0.086)	56
3 (0.099)	48
4 (0.112)	40
5 (0.125)	40
6 (0.138)	32
8 (0.164)	32
10 (0.190)	24
12 (0.216)	24
$\frac{1}{4}$	20
$\frac{5}{16}$	18
$\frac{3}{8}$	16
$\frac{7}{16}$	14
$\frac{1}{2}$	13
$\frac{1}{2}$	12
$\frac{9}{16}$	12
$\frac{5}{8}$	11
$\frac{3}{4}$	10
$\frac{7}{8}$	9
1	8
$1\frac{1}{8}$	7
$1\frac{1}{4}$	7
$1\frac{3}{8}$	6
$1\frac{1}{2}$	6
$1\frac{3}{4}$	5
2	$4\frac{1}{2}$
$2\frac{1}{4}$	$4\frac{1}{2}$
$2\frac{1}{2}$	4
$2\frac{3}{4}$	4
3	4
$3\frac{1}{4}$	4
$3\frac{1}{2}$	4
$3\frac{3}{4}$	4
4	4

TABLE F-2
UNF

Diameter	Threads per inch, P
0 (0.060)	80
1 (0.073)	72
2 (0.086)	64
3 (0.099)	56
4 (0.112)	48
5 (0.125)	44
6 (0.138)	40
8 (0.164)	36
10 (0.190)	32
12 (0.216)	28
$\frac{1}{4}$	28
$\frac{5}{16}$	24
$\frac{3}{8}$	24
$\frac{7}{16}$	20
$\frac{1}{2}$	20
$\frac{9}{16}$	18
$\frac{5}{8}$	18
$\frac{3}{4}$	16
$\frac{7}{8}$	14
1	12
$1\frac{1}{8}$	12
$1\frac{1}{4}$	12
$1\frac{3}{8}$	12
$1\frac{1}{2}$	12

TABLE F-3
Series 8

Diameter	Threads per inch, P
$1\frac{1}{8}$	8
$1\frac{1}{4}$	8
$1\frac{3}{8}$	8
$1\frac{1}{2}$	8
$1\frac{5}{8}$	8
$1\frac{3}{4}$	8
$1\frac{7}{8}$	8
2	8
$2\frac{1}{8}$	8
$2\frac{1}{4}$	8
$2\frac{1}{2}$	8
$2\frac{3}{4}$	8
3	8
$3\frac{1}{4}$	8
$3\frac{1}{2}$	8
$3\frac{3}{4}$	8
4	8
$4\frac{1}{4}$	8
$4\frac{1}{2}$	8
$4\frac{3}{4}$	8
5	8
$5\frac{1}{4}$	8
$5\frac{1}{2}$	8
$5\frac{3}{4}$	8
6	8

TABLE F-4
Series 12

Diameter	Threads per inch, P
$\frac{1}{2}$	12
$\frac{5}{8}$	12
$1\frac{1}{16}$	12
$\frac{3}{4}$	12
$1\frac{3}{16}$	12
$\frac{7}{8}$	12
$\frac{15}{16}$	12
$1\frac{1}{16}$	12
$1\frac{3}{16}$	12
$1\frac{5}{16}$	12
$1\frac{7}{16}$	12
$1\frac{5}{8}$	12
$1\frac{3}{4}$	12
$1\frac{7}{8}$	12
2	12
$2\frac{1}{8}$	12
$2\frac{1}{4}$	12
$2\frac{3}{8}$	12
$2\frac{1}{2}$	12
$2\frac{5}{8}$	12
$2\frac{3}{4}$	12
$2\frac{7}{8}$	12
3	12
$3\frac{1}{8}$	12
$3\frac{1}{4}$	12
$3\frac{3}{8}$	12
$3\frac{1}{2}$	12
$3\frac{5}{8}$	12
$3\frac{3}{4}$	12
$3\frac{7}{8}$	12
4	12
$4\frac{1}{4}$	12
$4\frac{1}{2}$	12
$4\frac{3}{4}$	12
5	12
$5\frac{1}{4}$	12
$5\frac{1}{2}$	12
$5\frac{3}{4}$	12
6	12

G

Standard Metric Threads

NOMINAL DIA (mm)	COARSE		FINE	
	PITCH	TAP	PITCH	TAP
1.6	0.35	1.25		
2	0.40	1.6		
2.5	0.45	2.05		
3	0.5	2.5		
4	0.7	3.3		
5	0.8	4.2		
6	1	5.0		
8	1.25	6.7	1	7.0
10	1.5	8.5	1.25	8.0
12	1.75	10.2	1.25	10.8
16	2	14	1.5	14.5
20	2.5	17.5	1.5	18.5

NOTE: OTHER SIZES ARE AVAILABLE
BUT THOSE LISTED ARE "PREFERRED"
FOR GENERAL USE

H

Standard Component Sizes

THESE DIMENSIONS ARE TYPICAL SIZES.
MANY OTHER SIZES ARE AVAILABLE

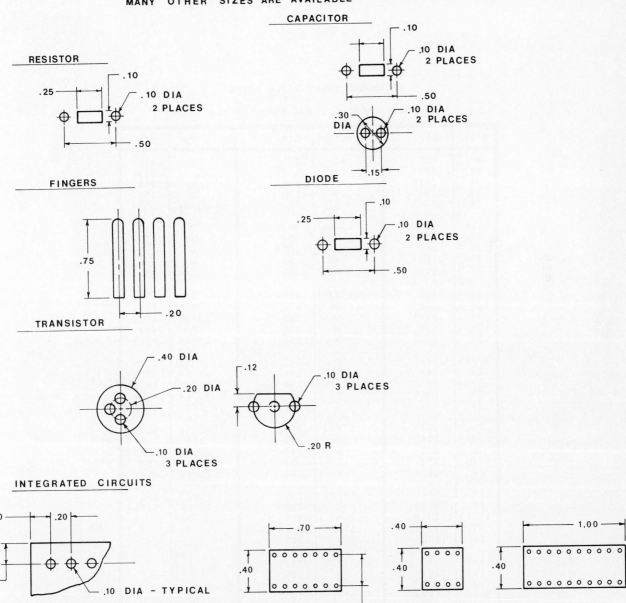

RESISTOR

CAPACITOR

FINGERS

DIODE

TRANSISTOR

INTEGRATED CIRCUITS

I

Standard Twist Drill Sizes

STANDARD TWIST DRILL SIZES

by Number, Letter, and Fraction (inches)

SIZE	DIA	SIZE	DIA	SIZE	DIA
40	.098	11	.191	R	.339
39	.0995	10	.1935	11/32	.3438
38	.1015	9	.196	S	.348
37	.104	8	.199	T	.358
36	.1065	7	.201	23/64	.3594
7/16	.1094	13/64	.2031	U	.368
35	.110	6	.204	3/8	.375
34	.111	5	.2055	V	.377
33	.113	4	.209	W	.386
32	.116	3	.213	25/64	.3906
31	.120	7/32	.2188	X	.397
1/8	.125	2	.221	Y	.404
30	.1285	1	.228	13/32	.4062
29	.136	A	.234	Z	.413
28	.1405	B	.238	27/64	.4219
9/64	.1406	C	.242	7/16	.4375
27	.144	D	.246	29/64	.4531
26	.147	1/4(E)	.250	15/32	.4688
25	.1495	F	.257	31/64	.4844
24	.152	G	.261	1/2	.5000
23	.154	17/64	.2656	9/16	.5625
5/32	.1562	H	.266	5/8	.625
22	.157	I	.272	11/16	.6875
21	.159	J	.277	3/4	.750
20	.161	K	.281	13/16	.8125
19	.166	9/32	.2812	7/8	.875
18	.1695	L	.290	15/16	.9375
11/64	.1719	M	.295		
17	.173	19/64	.2969		
16	.177	N	.302		
15	.180	5/16	.3125		
14	.182	O	.316		
13	.185	P	.323		
3/16	.1875	21/64	.3281		
12	.189	Q	.332		

STANDARD TWIST DRILL SIZES (millimeters)				
DIA	DIA	DIA	DIA	DIA
0.40	2.30	6.00	10.00	22.00
0.42	2.35	6.10	10.20	22.50
0.45	2.40	6.20	10.50	23.00
0.48	2.40	6.30	10.80	23.50
0.50	2.45	6.40	11.00	24.00
0.55	2.50	6.50	11.20	24.50
0.60	2.60	6.60	11.50	25.00
0.65	2.70	6.70	11.80	25.50
0.70	2.80	6.80	12.00	26.00
0.75	2.90	6.90	12.20	26.50
0.80	3.00	7.00	12.20	27.00
0.85	3.10	7.10	12.50	27.50
0.90	3.20	7.20	12.80	28.00
0.95	3.30	7.30	13.00	28.50
1.00	3.40	7.40	13.20	29.00
1.05	3.50	7.50	13.50	29.50
1.10	3.60	7.60	13.80	30.00
1.15	3.70	7.70	14.00	30.50
1.20	3.80	7.80	14.25	31.00
1.25	3.90	7.90	14.50	31.50
1.30	4.00	8.00	14.75	32.00
1.35	4.10	8.10	15.00	32.50
1.40	4.20	8.20	15.25	33.00
1.45	4.30	8.30	15.50	33.50
1.50	4.40	8.40	15.75	34.00
1.55	4.50	8.50	16.00	34.50
1.60	4.60	8.60	16.25	35.00
1.65	4.70	8.70	16.50	35.50
1.70	4.80	8.80	16.75	36.00
1.75	4.90	8.90	17.00	36.50
1.80	5.00	9.00	17.25	37.00
1.85	5.10	9.10	17.50	37.50
1.90	5.20	9.20	18.00	38.00
1.95	5.30	9.30	18.50	40.00
2.00	5.40	9.40	19.00	42.00
2.05	5.50	9.50	19.50	44.00
2.10	5.60	9.60	20.00	46.00
2.15	5.70	9.70	20.50	48.00
2.20	5.80	9.80	21.00	50.00
2.25	5.90	9.90	21.50	

NOTES FOR TWIST DRILL SIZES - INCHES

1. This is only a partial list of standard drill sizes.
2. Whenever possible, specify holes sizes that correspond to standard drill sizes.
3. Drill sizes are available in 1/64 increments between .5000 and 1.2500.
4. Drill sizes are available in 1/32 increments between 1.2500 and 1.5000.
5. Drill sizes are available in 1/16 increments between 1.5000 and 2.0000.

Index